T0314126

AN INTRODUCTORY GUIDE TO VALVE SELECTION

Other titles in this series:

An Introductory Guide to Valve Selection

Isolation, check, and diverter valves for the energy, process, oil, and gas industries

E. SMITH
and
B. E. VIVIAN

Series Editor
Roger C. Baker

Mechanical Engineering Publications Limited, London

First published 1995
Second Edition 1996

ISBN 0 85298 914 8

A CIP catalogue record for this book is available from the British Library.

Typeset by Paston Press Ltd, Loddon, Norfolk

MIX
Paper from
responsible sources
FSC
www.fsc.org   FSC® C013604

SERIES EDITOR'S FOREWORD

As an engineer I have often felt the need for introductory guides to aspects of engineering outside my own area of knowledge. MEP welcomed the concept of an introductory series to follow on from my own book on flow measurement. We hope that the series will provide engineers with an easily accessible set of books on common and not-so-common areas of engineering. Each author will bring a different style to his subject, but some valued features of the original volume, such as conciseness and the emphasis of certain sections by shading, have been retained. The initial volumes are biased towards fluids, but we hope to broaden the scope in later volumes.

The series is designed to be suitable for practising engineers and technicians in industry, for design engineers and those responsible for specifying plant, for engineering consultants who may need to set their specialist knowledge within a wider engineering context, and for teachers, researchers, and students. Each book will give a clear introductory explanation of the technology to allow the reader to assess commercial literature, to follow up more advanced technical books, and to have more confidence in dealing with those who claim an expertise in the subject.

Edwin Smith has adapted this fourth volume in the series from a BP 'Recommended Practice' manual, which he wrote with the help of Barry Vivian, his distinguished and long-serving predecessor as BP's valve specialist. Mr Smith's previous experience has included over fifteen years in the valve industry and several years in nuclear power project work. He is, therefore, well placed to undertake the task of adapting this manual from BP's immense experience in one of the most demanding industries for engineering equipment. The result is not only an introductory guide to valves but an immensely useful handbook even for those with knowledge of the subject. It is enhanced by extensive application tables, lists of relevant standards, and a very useful glossary. It therefore adds a most valuable fund of information to this growing series, and I hope it will be welcomed by engineers and find a valued place on their shelves of well-thumbed reference books.

I would particularly like to acknowledge the encouragement of Mick Spencer, the Managing Editor, to develop the series, and the enthusiasm and help of Louise Oldham, the Coordinating Editor who has been responsible for this volume. We all hope that the series will find a welcome with engineers and we shall value reactions and any suggestions for further volumes in the series.

Roger C. Baker
St Albans

CONTENTS

AUTHORS' PREFACE

This book aims to provide guidance on the choice of common types of isolating (block), check and diverter valves for the energy, process, oil, and gas industries. It is applicable to both onshore and offshore locations, including sub-sea applications. Whilst the experience on which the text is based derives from these industries and applications, the authors hope that readers with a more general interest in valves will also find the book to be of value.

In Chapter 1 the main types and features of valves are described and the following chapter reviews the wide range of service conditions in which they are used.

Chapter 3 reviews the different types of block and diverter valves in detail, whilst Chapter 4 deals with the requirements for operation and isolation. Chapter 5 describes the various types of check valve and Chapter 6 discusses valves for various special applications which are commonly met with. Wellhead gate valves for the petroleum industry are dealt with in a separate chapter (7).

Metallic and non-metallic materials are reviewed in Chapter 8 – application and chemical resistance tables being provided to aid selection – and the final chapter presents simple formulae for calculation of head loss and surge potential.

Valve selection tables are provided in Appendix A and the remaining three appendices list abbreviations, relevant standards and a glossary of terms, respectively.

E. Smith
B. E. Vivian

ACKNOWLEDGEMENTS

The authors gratefully acknowledge all those organizations and companies who have given permission for the use of illustrations.

The illustrations provided by the British Valve and Actuator Manufacturers' Association are taken from their publication *Valves and Actuators from Britain* which is a complete guide to the products of their members.

Thanks are also due to Mr D. G. King who provided much assistance during the preparation of the in-house document on which this book is based.

ORIGIN

This text derives from a BP Recommended Practice which was based on experience of valves in service and extensive testing carried out with the co-operation of valve manufacturers. MEP, BP, the authors, and the editor do not accept any responsibility for the consequences of implementing the recommendations or using the information contained in this publication.

CHAPTER 1

Basic Valve Types and Function

1.1 INTRODUCTION

The modern processing plant contains hundreds, and often thousands, of isolating or block valves in a wide range of sizes, pressure ratings, materials, and types. If they are to perform acceptably during a long life, careful attention needs to be given to their selection and specification; the universal valve, suitable for all applications, does not yet exist.

Valve selection is made from a knowledge of:

(a) required function
(b) service characteristics

The selection of valves requires the consideration of many factors in addition to the guidelines given here, and past experience of particular applications should always be taken into consideration. Many of the factors involved can be simplified by an early evaluation of valve requirements and preparation of procurement specifications that adequately define them. This approach can be of benefit in modifying existing plant, is of considerable importance on new projects, and may be of overriding importance where valve development is required for special applications.

General technical factors that must be taken into account include:

(a) weight   (d) reliability
(b) size   (e) operability
(c) ease of maintenance

There will also be commercial factors, such as cost and required delivery, which can influence valve choice in many circumstances, particularly for less severe services.

1

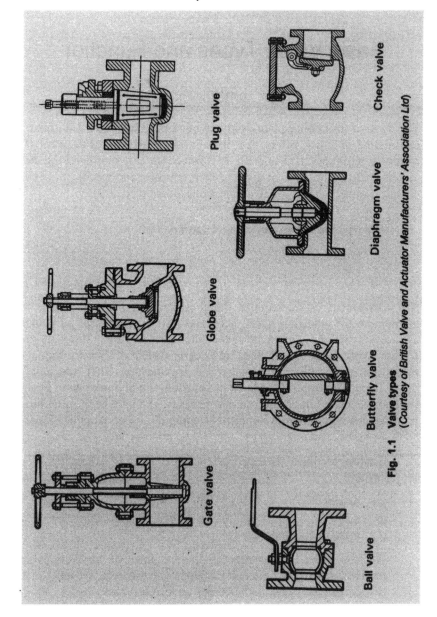

Fig. 1.1   Valve types
(*Courtesy of British Valve and Actuator Manufacturers' Association Ltd*)

1.2 VALVE TYPES

The basic types of valve covered by this text are illustrated in Fig. 1.1 (opposite). It will be seen that the path taken by the fluid when flowing through a fully open valve varies in complexity, depending on the valve type. The energy dissipated in friction or noise will be greater the more tortuous the path, and this may be a consideration in valve selection.

The configuration of the flow path through a valve and the method used to restrict or arrest flow determine a valve's characteristics and influence the selection of a particular type of valve for a particular function (Fig. 1.2).

The method of closure can be important in valve selection, especially when considering isolation of dirty fluids, and valves may be grouped according to the method by which the obturator or closure member (gate, ball, disk, plug, etc.) moves relative to the seat.

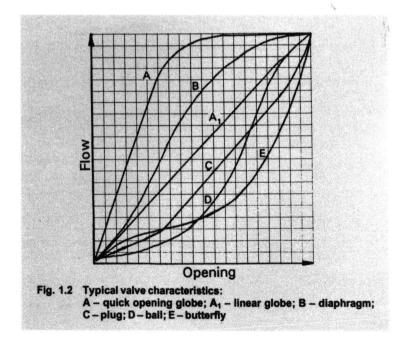

Fig. 1.2 Typical valve characteristics:
A – quick opening globe; A₁ – linear globe; B – diaphragm;
C – plug; D – ball; E – butterfly

> The movement relative to the valve seat may be sliding, closing, or flexing and the path of travel may be linear or rotary. Valves are often described as having linear action (or multiple turn if screw-operated) or rotary action (more commonly, quarter turn).

The different methods of restricting or arresting flow are as follows:

(*a*)   *Sliding method*
The closure member slides across the valve seat face to open or close the valve.

Linear action valves using the sliding method are parallel gate valves.

Rotary action valves that use this method are ball valves and plug valves (Fig. 1.3).

(*b*)   *Closing method*
The closure member moves away from or towards the valve seat to open or close the valve. Closure is achieved either by abutment against the seat face or by projection into the seat orifice.

Linear action valves using the closing method are globe valves and lift check valves.

Rotary action valves using the closing method are butterfly valves and swing check valves (Fig. 1.4).

(*c*)   *Flexing method*
Opening or closing is achieved by flexure of a resilient membrane within the valve body.

Linear action valves using this method include diaphragm and pinch valves. The mechanical closure member is external to the fluid flow and external fluid pressure may also be used to flex the membrane (Fig. 1.5).

Rotary action valves using this method are relatively uncommon, but Iris valves, in which a flexible membrane of tubular shape is rotated into a conical shape for closure, are an example.

Combinations of the above methods with linear/rotary action are not uncommon. The widely-used wedge gate valve is an example of a linear action valve which appears to employ the sliding method but actually uses the closing method to wedge the gate into the tapered seats. The lifting

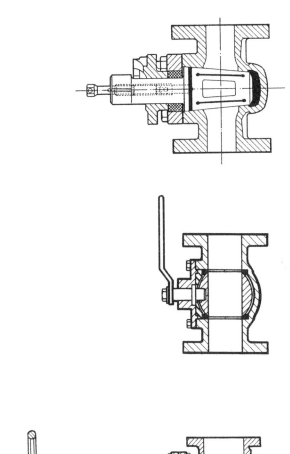

Fig. 1.3 Valves which utilize the sliding method for stopping flow
(Courtesy of British Valve and Actuator Manufacturers' Association Ltd)

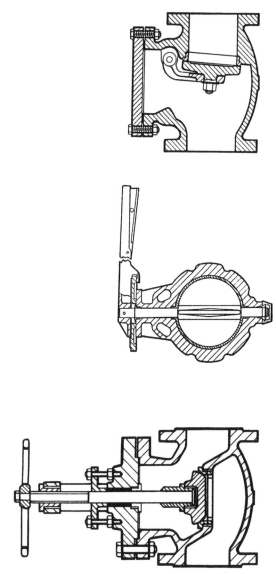

Fig. 1.4   Valves which utilize the closing method for stopping flow
(Courtesy of British Valve and Actuator Manufacturers' Association Ltd)

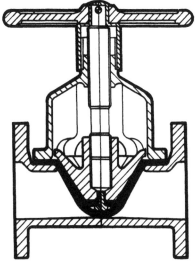

Fig. 1.5   Diaphragm valves, use the 'flexing' method
(*Courtesy of British Valve and Actuator Manufacturers' Association Ltd*)

taper plug (wedge plug) is an example of combining linear and rotary action with the closing method.

Typical valve types used for the basic operating functions are given in the following table.

Valve function	Typical valve types
Isolation	Gate
	Ball
	Butterfly
	Plug
	Diaphragm/pinch
	Globe
Flow diversion	Plug
	Ball
	Globe
Prevention of flow reversal	Swing check
	Lift check
	Diaphragm check
	Axial flow check

CHAPTER 2

Service Characteristics

2.1 GENERAL

Appropriate valve selection is dependent on complete knowledge of the service characteristics.

(a) *Fluid type*
The fluid being handled should be classified as liquid, gas, two-phase mixture, steam, slurry, or solids.

(b) *Fluid characteristics*
One or more of the following characteristics may be attributed to the fluid: clean; dirty (including abrasive); containing large, suspended solids; liable to solidification; viscous; corrosive; flammable; fouling or scaling; of a searching nature.

(c) *Pressure, temperature limitations, and chemical resistance*
Valves are normally allocated a rating according to the maximum operating pressure and temperature, which is compatible with the rating of the connected piping system or flanges. The operating temperature limits the materials which may be used in the valve construction, particularly for trims, seals, linings, or lubricants.

Materials may also be limited by the pressure, fluid concentration, and condition. Metallurgical advice should always be sought where doubt exists.

Irrespective of chemical resistance properties, cast iron, copper alloy or plastic valves should not be used on hydrocarbon, toxic, or other hazardous service.

(d) *Operation and maintenance requirements*
Operational and maintenance requirements can influence selection and design. Consideration should be

given to:
- fire resistance;
- ease and speed of operation;
- leak tightness (internal and external);
- maintainability;
- weights and dimensions (construction, handling etc);
- storage and commissioning;
- location (e.g., sea bed valves);
- line clearing requirements (e.g., ability to pass; cleaning pigs).

Some of these requirements are discussed in more detail below.

2.2 FLUID CHARACTERISTICS

The characteristics and condition of fluids and slurries require careful identification since these are often the most significant factors in selecting the correct type of valve. Clean fluids generally permit a wide choice of valve types; for dirty fluids the choice is often restricted.

Other factors to be considered include: the hazardous nature of the fluid (flammability, toxicity, and searching nature) which may impose limits on permissible leakage to atmosphere or past valve seats; corrosivity; viscosity and tendency to solidify.

2.3 CLEAN SERVICE

Clean service is a term used to identify fluids which are free from solids or contaminants.

Fluids normally defined as 'clean' include instrument air, nitrogen, potable water, treated (demineralized) water, steam, lube oil, diesel oil, and many chemicals including dosing chemicals used for injection into fluid systems for which there are no special clean requirements. Valves for fluids such as oxygen, hydrogen peroxide, and sometimes treated water or lube oil require special attention to cleanliness. Thorough degreasing and clean room assembly will often be necessary. Valves for potable water must meet national regulatory requirements.

Some hydrocarbon services may also be defined as clean, subject to the consideration of conditions at each stage of processing.

Clean services are generally less damaging to valves, resulting in long-term performance and reliability. Selection from a wide range of valve types may be possible, thus allowing greater freedom of choice.

If the service is basically clean, attention should be given to protecting valves during construction and commissioning where conditions detrimental to long valve life often exist. This may involve the removal of valves or delicate internal parts until flushing, pigging, and drying operations are complete and the piping system has been cleaned.

2.4   DIRTY SERVICE

Dirty service is a general term used to identify fluids with suspended solids that may seriously impair the performance of valves unless the correct type is selected. This type of service is often of major significance since many valves are very sensitive to the presence of solids. Dirty service may be further classified as abrasive or sandy.

2.5   ABRASIVE SERVICE

Abrasive service is a term used to characterize fluids containing abrasive particulate such as pipe rust, scale, welding slag, sand, and grit. These materials can damage seating surfaces and clog working clearances in valves, often resulting in excessive operating force requirements, sticking, jamming, and unacceptable leakage through the valve. Such damage may be caused by particulate in quite low concentrations and of small size – typically 10 microns.

Abrasive conditions are commonly found during construction, continuing into production. Valves can be irreversibly damaged and may require early overhaul, unless adequate preventative measures are taken.

Where abrasive conditions continue during production or protection during commissioning is not possible, a valve suitable for dirty service should be selected. Typical conditions include: naturally-occurring particulate in the process fluid (e.g., sand from production wells); corrosion products from pipe surfaces (which could be caused by the change in the nature of the fluid, the injection of dosing chemical during service, or the release of pipe scale following drying out of piping in gas systems); and abrasive compounds formed during chemical processes.

2.6 SANDY SERVICE

Sandy service is a term identifying severe abrasive and erosive conditions and is used in oilfield production to identify the inclusion of formation sand with reservoir crude oil or gas.

Valves for this service are often required to have their performance qualified by means of the sand slurry test specified in API 14D.

2.7 FOULING SERVICE

Fouling or scaling services are general terms used to identify liquids or elements of liquids that form a deposit on surfaces. Such deposits may vary widely in nature, with varying hardness, strength of adhesion, and rates of build up. Valves for these services require careful selection particularly where thick, hard, strongly adhesive coatings occur. The temperature of the fluid may be a vital factor and in some cases valves may need to be steam jacketed or trace heated. Purging (e.g., with steam or clean product) is often employed.

2.8 SLURRY SERVICE

Slurry service is a general term used to define liquids with substantial solids in suspension. Often the product is the solid and the fluid is primarily the means of transportation, e.g., coal slurries and catalyst services. Slurries vary widely in nature and concentration of solids. Hard solids in high concentration can cause severe abrasion, erosion, and clogging of components. Soft, non-abrasive solids can cause clogging.

Differential expansion at elevated temperatures also requires careful consideration in valve designs for slurry service.

2.9 SOLIDS

There are many other services where solids may be present in the form of hard granules, crystals, soft fibres or powders. The transporting medium may be liquid or gas. Air of fluidized bed systems may be used for some particulates. Specialized valves are available for many of these services but development work may sometimes be necessary.

2.10 HAZARDOUS SERVICE

Where the term 'hazardous service' is used in this document, this will usually be defined for each specific case e.g., using the guidance provided by the European Pressure Equipment Directive. However, it will usually include the following:

(a) liquids above their auto-ignition temperature (AIT), or 210°C if the AIT is not known;

(b) flammable liquids flashing on leakage to form a substantial vapour cloud; this will include LPG, LNG, NGL condensate, and some others;

(c) fluids liable to cause a hazard by blockage due to hydrate formation or solids deposition;

(d) toxic substances, (e.g., chlorine, hydrofluoric acid, hydrogen sulphide, CO, phenol etc.);

(e) hydrogen service – defined as service in contact with hydrogen or gaseous mixtures containing hydrogen in which the partial pressure of hydrogen is 5 bar (abs), (72.5 psia) or more;

(f) flammable fluids at Class 900 flange rating and above;

(g) highly corrosive fluids such as acids and caustic alkalis;

(h) scalding fluids e.g., hot boiler feed water, steam above class 300.

2.11 FLAMMABLE SERVICE

This includes any fluid with an auto-ignition temperature (AIT) above 210°C, and fluids which will flash off an inflammable vapour cloud.

2.12 SEARCHING SERVICE

Searching service is a term used to identify fluids that require special attention in valve design and manufacture to prevent leakage through pressure-containing components (body-bonnet joints, glands, etc.) and through seats and seals.

Gases of low molecular weight such as hydrogen and helium and liquids of low viscosity such as Dowtherm are typical examples.

2.13 SOLIDIFYING SERVICE

Solidifying service is a general term used to identify fluids that will change from liquid to solid unless maintained at the correct conditions of temperature, pressure, and flow. It is a term generally associated with fluids such as liquid sulphur and phthalic anhydride where valves of steam jacketed design may be required, or heavy fuel oil, where valves often require trace heating to maintain temperature and operability. In certain chemical processes polymerization may occur, thereby blocking cavities and preventing valve operation.

2.14 CORROSIVE SERVICE

Corrosive service is a term generally used to identify fluids containing corrosive constituents that, depending on concentration, pressure, and temperature, may cause corrosion of metallic components.

Corrosive fluids include sulphuric acid, acetic acid, hydrofluoric acid (HFA), wet acid gas (wet CO_2), wet sour gas (wet H_2S), and chlorides. Many chemicals are highly corrosive including concentrations of some corrosion inhibitors.

Both dirty and clean services may contain corrosive fluids, e.g., a dosing chemical service with corrosion inhibitor could be nominally defined as clean.

The choice of suitable corrosion-resistant materials for valve wetted parts is necessary to avoid corrosion, which could impair the integrity or performance of the valve.

Types of corrosion that need to be considered when selecting valve materials and designs include the following:

(a) acid corrosion resulting in general wastage (typical with wet CO_2);
(b) crevice corrosion;
(c) galvanic corrosion between dissimilar materials;
(d) pitting corrosion (e.g., austenitic stainless steel in seawater);
(e) stress corrosion cracking of components (typical with wet H_2S and chlorides depending on concentration, pressure, and temperature);

Materials for sour (H_2S) service are required to conform with the requirements of NACE standard MR–01–75.

2.15   VISCOUS SERVICE

Viscous service is a term that generally identifies a wide range of fluids with pronounced thickness and adhesive properties that, for the range of operating conditions (pressure, temperature, and flow) may require high operating torques and cause a sluggish response, as a result of which seating is affected. Fluids include high viscosity oils (lube and heavy fuel oil) and non-Newtonian fluids e.g., waxy crude, gels, and pastes.

The choice of valves for viscous service can vary, depending on fluid properties. Special attention should be given to check valves where sluggish response may cause operating difficulties and even hazardous conditions.

2.16   VACUUM SERVICE

Vacuum service is a term used to identify systems where the pressure is permanently or intermittently below atmospheric. Valve selection should pay particular attention to the reverse sealing capability of glands etc.

In the case of systems which have the potential to create an unwanted vacuum (e.g., condensing, tanks etc.) it is often necessary to fit a vacuum breaker valve which functions in such a way as to admit air automatically whenever a vacuum occurs and so balance the barometric pressure.

2.17 FIRE HAZARD

Certain areas of an installation may be classified as presenting a special fire risk. This will usually require the fire testing of soft seated valves and the fire protection of all types of valves.

Valves in special fire risk areas will normally be identified as being on (a) critical or (b) non-critical duty.

(a) *Valves on critical duty* that are required to remain operable during any fire must be capable of remote operation from outside the fire risk area. Such valves are anticipated to be few in number on conventional onshore installations, but will probably be much more numerous on compact plants such as certain modular or offshore installations. The complete valve assembly including motor, actuator, and cabling will require protection against fire.

Valves on critical duty that are required to provide a seal during or after a fire will have either metal-to-metal (or non-decomposing) primary seats, seals etc. or, if fitted with soft primary seats, will be covered by fire test certificates in accordance with one of the recognised standards.

(b) For *valves on non-critical duties,* fire testing and associated documentation will not usually be a requirement.

CHAPTER 3

Valve Types for Isolation (Block) Duty

3.1 GENERAL

Isolation or block valves are used for starting and stopping flow and are generally selected to provide the following:

(a) Low resistance (pressure drop) to flow by means of a straight through flow configuration which also facilitates line clearing. (In smaller sizes, and where flow is not a requirement, the positive sealing attributes of the globe valve are often utilized.)

(b) Shut-off with flow or pressure from either direction i.e., bi-directional sealing.

Block valves are the most widely used valve type. Operation is normally by manual intervention either directly or via powered actuators.

Many types of valves are used, including the following.

Ball valves	– floating ball/trunnion mounted ball
Butterfly valves	– 'high performance'/rubber lined
Gate valves	– wedge/parallel slab gate/parallel slide
Globe valves	– straight/angle/oblique/needle/ piston/stop and check
Plug valves	– taper/parallel; lubricated/non- lubricated; lined/sleeved
Diaphragm valves	– weir/full flow/pinch valves

3.2 BALL VALVES

These are low torque, quarter turn rotary action valves with low resistance to flow and are suitable for many on–off utility and process services. They have a straight through configuration typical of the sliding method of closure. There are several designs including the floating, or seat-supported ball (Fig. 3.1), eccentric ball (Fig. 3.2) and trunnion mounted ball (Fig. 3.3) types. Most designs are double seated but there are some special single seated designs (Fig. 3.2).

The majority of valves have soft seats, usually PTFE or nylon, which can provide bubble tight sealing but which limit the maximum working temperature and make the valve unsuitable for abrasive service (where lips of soft seals are easily damaged by hard particles). Graphite seats are also available. Metal-to-metal seated designs with hard faced balls and seats can, however, be purchased and these are suitable for abrasive service where the fluid is dirty or hard particles are present. Seat sealing will not be as good as is the case with soft seated designs owing to the

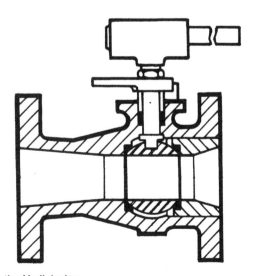

Fig. 3.1 'Floating' ball design
(*Courtesy of British Valve and Actuator Manufacturers' Association Ltd*)

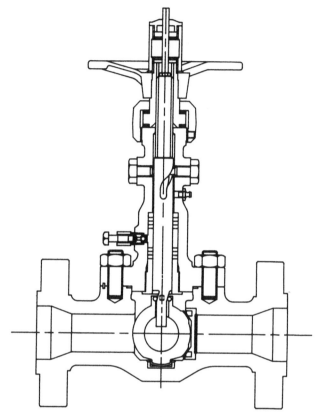

Fig. 3.2   'Eccentric' ball valve
(Courtesy of Orbit Valve plc)

difficulty of achieving completely spherical surfaces with hard materials. Although increasing seat flexibility can improve sealing, costs can escalate and a slab gate valve may offer an easier way of obtaining good seat tightness in such services.

Reduced bore valves are not recommended for very high velocities when fluids contain solids. Where scaling may take place on the ball surface, metal seats with a scraping action to clean the ball should be specified.

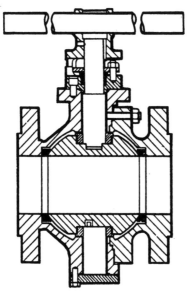

Fig. 3.3 Trunnion mounted design
(*Courtesy of British Valve and Actuator Manufacturers' Association Ltd*)

Solids can become trapped in body cavities during opening or closing operations and a cavity bleed or flushing facility may be required. Plated balls should be avoided in abrasive service and, where used elsewhere, consideration should be given to the ability of the base material to resist corrosion where water etc. may be present.

The body design features and ball support arrangements vary and can be defined as follows.

Body design features:
– one piece
– split-body

Method used to support ball:
– seat supported and floating ball type
– trunnion mounted type

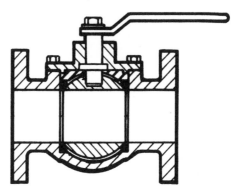

Fig. 3.4 Top entry ball valve
(*Courtesy of British Valve and Actuator Manufacturers' Association Ltd*)

Body designs

(*a*) *One piece*
    (i)   Axial (or end) entry (ball fitted through body ends and retained by a screwed or bolted ring (Fig. 3.1)).
    (ii)   Top entry (Fig. 3.4).
    (iii) Bottom entry.
    (iv) Sealed type (welded construction).

(*b*) *Split-body*
    (i)   Two piece – comprising body and body connector (Fig. 3.5).
    (ii)   Three piece – comprising body and two body end connectors incorporating the pipe flanges (Fig. 3.6). This may enable the body to be removed from the line leaving the body connectors attached to mating pipework. The body connectors usually retain the seats and sealing may be affected by the removal and replacement of the body. Has additional joints which could leak.

Ball support method

(*a*) *Seat supported and floating ball type* (Fig. 3.1)
These are designed to support the ball on two seats located on up-stream and down-stream sides. The upstream pressure presses the ball onto the downstream seat, compressing the seal and shutting off fluid flow. Some valves have pre-compressed seats facilitating a double block and bleed function; these should have features to allow

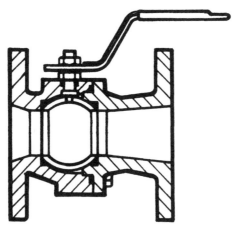

Fig. 3.5 Two piece ball valve
(*Courtesy of British Valve and Actuator Manufacturers' Association Ltd*)

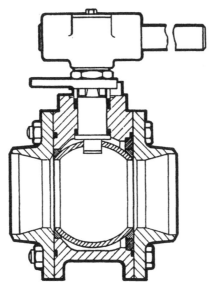

Fig. 3.6 Three piece ball valve
(*Courtesy of British Valve and Actuator Manufacturers' Association Ltd*)

relief of any body cavity over pressure. Seat-supported valves are generally used in small bore piping or low pressure clean service (classes 150–300). Larger sizes and higher pressures result in high seat loads and operating torques but are sometimes used in high temperature service for which the piston type seats of trunnion mounted valves would be unsuitable.

(b) *Trunnion mounted type* (Fig. 3.3)

These have the ball mounted on trunnions supported in body bearings above and below the ball. Sealing is achieved by spring loaded piston type seats which shut off flow when line pressure acts on the upstream seat (Fig. 3.7). This design provides automatic relief of cavity overpressure (Fig. 3.8). The standard design cannot be used for double block and bleed isolation duty but 'twin seal' designs are available which can, although they would seldom be first choice for such service. In this case, automatic cavity relief is forfeited and an external relief valve must be provided if this is a requirement.

Trunnion mounted designs are available in sizes above DN50 and are generally used for higher pressure service. They have lower operating torque requirements than seat supported types.

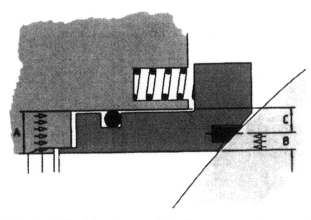

Fig. 3.7   Spring loaded, piston type seat. As line pressure increases, the seat differential area C = A − B creates a piston effect which forces the seat against the ball. The higher the line pressure, the greater the closing force (*Courtesy of Ring O Valve*)

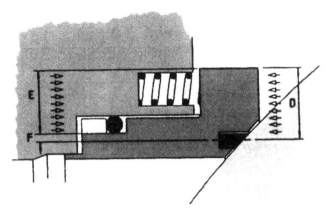

Fig. 3.8 Automatic relief of cavity overpressure. The piston action reverses if pressure in the body exceeds that in the line, allowing excess cavity pressure to be vented. The design must be such that F = D − E creates a greater force than the spring.
(*Courtesy of Ring O Valve*)

Where soft seats are specified, shrouded designs with a large contact area are recommended to minimize seat damage.

For high temperature service, graphite or metal seats should be specified.

Valves with chromium or nickel plated carbon steel balls should not be used in service where water is present since plating is porous and corrosion is likely to result. For abrasive service, balls and seats faced with tungsten carbide, Stellite 6, or similar will generally give best results. The material and method of deposition must, however, be carefully controlled.

Welding ends are usually only specified for sizes less than DN 50, and for large, top entry designs used in pipeline applications. The carbon content of socket or butt-weld components should be limited to 0.25 percent. Small, soft-seated socket or butt weld end valves should be provided with stub pipes welded in place by the manufacturer prior to valve assembly, the valves having an overall length of 400 mm or so to prevent damage to the seat during welding into the line. Flange valve sizes less than DN 50 are normally only used on equipment connections, tank nozzles, and header branches.

Where class 150 short pattern valve sizes DN 300 and DN 400 are specified the ball may protrude beyond the body end flange faces when the valve is closed. Such valves cannot be used where spading is envisaged, cannot be easily removed when in the closed position, and should generally be avoided.

Where specified with soft seats for use in a fire risk area valves should be of a fire tested design certified to a recognized standard and fitted with an anti-static device. For important isolation duties in fire risk areas, external protection may be considered.

Pressure/temperature ratings should normally comply with ANSI B16.34 and the requirements of BS 5351 or API 6D. It should be noted that, although ratings go down to −29°C, this relates to the body material; at low temperatures, the suitability of any polymers or elastomers should be investigated and steels should be demonstrated to possess adequate toughness. Upper temperature ratings will usually also be limited by the presence of polymers and elastomers and the manufacturer's advice should be sought in such cases.

Valves are available which provide sealant injection to seal damaged seating surfaces but this is not always effective and the seal will generally not survive more than one operation.

Where rapid closure of the valve could cause water hammer, gear operated valves should be used.

3.3   BUTTERFLY VALVES

These are low torque, quarter turn, rotary action valves with a straight through flow configuration in which a disk is turned in axial trunnion bearings onto a single seat and intrudes into the flow passage in the fully open position (Fig. 3.9). They are of compact design and may be obtained with or without flanges and linings. Seating arrangements may be soft (use of body lining, trapped 'O' ring etc.) or metal-to-metal.

Butterfly valves generally do not provide the lowest resistance to flow, especially as the size decreases, due to the increasing intrusion of the disk into the flow path. The configuration also imposes pressure limitations and class 300 rating is the normal upper limit with some special designs extending this to class 600. The closing method can provide good shut off

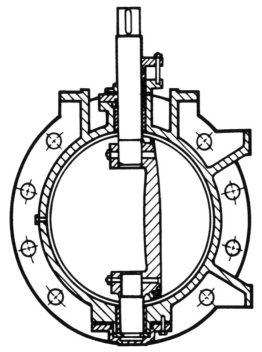

Fig. 3.9   Butterfly valve
(Courtesy of British Valve and Actuator Manufacturers' Association Ltd)

and various degrees of seat leak tightness, dependent on the design. Butterfly valves are generally suitable for bi-directional sealing but some designs have a preferred direction of flow.

Butterfly valves have the following disadvantages:

(a)   the line cannot be pigged;

(b)   they create higher pressure drop than full bore gate or ball valves;

(c)   they have to be withdrawn from the line for main-tenance;

(d)   double block and bleed facility is not available unless two valves are installed in series.

Butterfly valves are available in two basic types based on the method of mounting the disk.

> **Butterfly valve types:**
>
> – **conventional – shaft passes through the disk axis;**
> – **'high performance' – with offset disk/seats for higher pressure.**

(*a*)   *Conventional type*

The shaft passes through the disk axis and tight shut-off is not normally achievable except in the case of some rubber-lined designs. Applications include the following.

(i)   Low pressure (class 150 max) isolation. Maximum shut-off pressure: generally 14 bar (refer to manufacturers' catalogues). Maximum operating temperature 120°C, dependent on resilient lining. (Refer to manufacturers' catalogues).

(ii)   Shut-off with relatively high seat leakage, up to 425°C for metal seated valves.

(iii)   Flow control service. (Note: when used for control, butterfly valves exhibit high pressure recovery downstream and are thus susceptible to generating cavitation in liquid service).

(iv)   Non-demanding corrosive service.

(v)   Large flows of gases, liquids, slurries, liquids with solids in suspension.

Body patterns available are:

(i)   double flanged (Fig. 3.9) – this has flanged ends for connection to pipe flanges by individual bolting. Flanged valves are recommended in larger sizes to facilitate handling.

(ii)   wafer (Fig. 3.10) – this is primarily intended for clamping between pipe flanges using through bolting with a body of one of the following patterns:
        – single flange or lug type which incorporates through drilled or tapped holes;

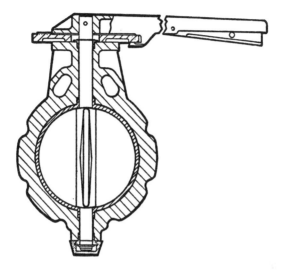

Fig. 3.10 Wafer-style butterfly valve
(*Courtesy of British Valve and Actuator Manufacturers' Association Ltd*)

- – flangeless which is fitted inside the bolt circle;
- – 'U' section which has flanges, but is generally not assumed suitable for individual bolting of each flange to the pipework.
(b) *'High performance' type* (Fig. 3.11)
This refers to valves with offset disk/seats which are suitable for higher pressures. The term does not exclude conventional valves from having a high performance within their pressure/temperature range. Applications include the following:

(i)   low to medium pressure (class 150 to class 600); (note that there are limited suppliers of class 600 valves).

(ii)   differential pressure can be to full flange rating except where restricted by resilient seat material;

(iii) tight shut-off at temperatures up to 260°C using resilient seats;

(iv)   shut-off with some seat leakage at temperatures up to 538°C for metal seated valves;

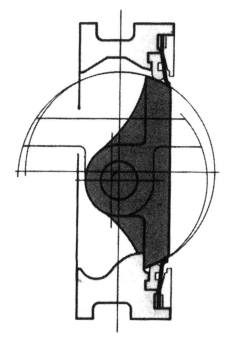

Fig. 3.11   'High performance' offset disk butterfly valve
(*Courtesy of Charles Winn (Valves) Ltd*)

(v)   on–off or control service (see above);
(vi)   large flows of gases, liquids, slurries, liquids with solids in sus-
        pension;
(vii)   fire safe application;
(viii) cryogenic service.

Body patterns available are normally restricted to single flange/lug
type or flangeless wafer type.

Conventional butterfly valves are usually supplied with iron or steel
bodies. Iron valves should not be used for process duties, hazardous
service, or where freezing is a possibility. The valve is usually installed
with the shaft passing through the vertical axis of the disk. For tight
closure the disk/seat interface must be suitably designed, particularly

around the shaft. Tight closure can be achieved by mounting the disk eccentrically on the stem/shaft providing an uninterrupted 360 degrees seal.

Lined valves can perform well on gritty or abrasive services.

Vulcanized linings are difficult to renew, and are not recommended.

'High performance' butterfly valves are usually supplied with steel or bronze bodies, and have the shaft mounted eccentrically on the disk.

Single flange (lug) type valves with replaceable seats may not be suitable for dead-end service without the use of a downstream companion or blind flange.

A wafer type butterfly valve in which the resilient seat is extended to serve also as a line gasket should only be installed between weld neck or socket weld pipe flanges. Slip-on or threaded flanges may not provide an adequate seal.

Valves in which the gasket contact area is reduced by counterbored or countersunk holes for retaining screws (used to secure seat ring assemblies in the valve body) should be avoided if possible. When specified, they should be used only with the gasket manufacturer's recommended gasket, contact area, and surface finish. Spiral wound gaskets should not be used on such valves. Also to be treated with caution are valves which rely on compression between the pipe flanges to retain or seal the seat.

During operation the distribution of static fluid pressure on the disk may produce a strong closing torque and larger size valves should be equipped with self-locking gearing or other substantial restraint on uncontrolled shaft rotation.

On liquid service, manually operated valves located such that rapid closure could produce water hammer should be gear operated. Alternatively, if suitable, a gate valve may be specified.

The user should ensure that the disk, when fully or partly open, will not foul adjacent valves, fittings, or connected pipework, particularly when used in cement or rubber lined piping systems.

Flangeless wafer-type valves with long exposed bolts, may cause leakage when subjected to high or low temperatures (or when exposed to fire) due to expansion of the bolts and consequent unloading of the joint gaskets. Where such valves are used for flammable or toxic service a light gauge stainless steel shroud should be wrapped around the valve and exposed bolts. Alternatively, lug-type valves, which protect the bolts, may be considered. In order to limit the potential for leakage in such conditions bolting should be kept as short as possible.

Wafer-type valves should not be used in cases in which it is necessary to dismantle a pipeline leaving the valve at the end of the pressurized line. Valves having tapped holes should not be used in this service as bolts may corrode in the body and become difficult to remove.

Note that valves procured to different standards may not be interchangeable because of differing face-to-face dimensions.

3.4 GATE VALVES

Gate valves are used for on/off operation on hydrocarbon, general process and utilities service for all temperature ranges. They have straight through configurations which are typical of the sliding method of closure.

Gate valve types include the following:

(a) wedge
(b) parallel double disk (expanding wedge)
(c) parallel slab/through conduit
(d) parallel slide
(e) knife-edge
(f) venturi (variation on (a) and (d))
(g) compact (variation on (a))

Gate valves should not be used for the following applications:

(a) installation in horizontal lines transporting heavy or abrasive slurries where sediment may become trapped in the pocket below the valve seat, preventing closure (except for reverse acting conduit or knife-edged types);
(b) throttling duties generally, as erosion of seats and disk may occur, causing leakage; gate valves may, however, be used for control valve bypass duty.

Gate valves have the following stem arrangements:

(a)   inside screw; rising or non-rising stem;
(b)   outside screw; rising or non-rising handwheel.

An outside screw valve is easier to maintain, allowing access for lubrication of the thread. In marine environments the rising stem and threads should be protected against corrosion, or corrosion-resistant materials should be specified.

Where headroom is limited, a non-rising stem type valve may be specified (Fig. 3.12). As the stem thread is within the body and is exposed to the line fluid, this type is unsuitable for corrosive or slurry service where excessive wear may occur on threads, or for high temperature applications where expansion and contraction may cause thread binding. BS 5352 restricts carbon steel valves of this type to a maximum temperature of 425°C.

When quick and frequent operation is necessary, and taking account of water hammer, ball valves, or in some instances plug valves which have a quarter turn operation, are preferred rather than the relatively slower operating gate valve. Note, however, that where gear operators are employed with such designs, operating times are comparable with those of the gate valve.

3.4.1   Wedge gate valve

Steel wedge gate valves (Fig. 3.13) are classified by wedge type: plain solid wedge; flexible solid wedge; split wedge. A flexible solid wedge may more easily accommodate misaligned seats and minimize galling of sealing surfaces. A plain solid wedge may be more difficult to grind to an accurate fit.

Solid wedge gate valves (Fig. 3.13(b)) are good general block valves offering a good sealing capability with low pressure drop. A 100 percent shut-off capability cannot always be relied upon, however, and slight leakage may occur with variations in temperature and pressure after being in service for some time. For hydrogen service, double (two) wedge gate valves should be used for shut-off applications. Extended bonnets are available for cryogenic service.

Wedge gate valves are prone to 'thermal wedging' when subjected to temperature changes after closure. In these and similar conditions, where the valve body may deform following a change in process conditions, a flexible wedge type valve (Fig. 3.14(a)) or a split wedge valve (Fig. 3.14(b))

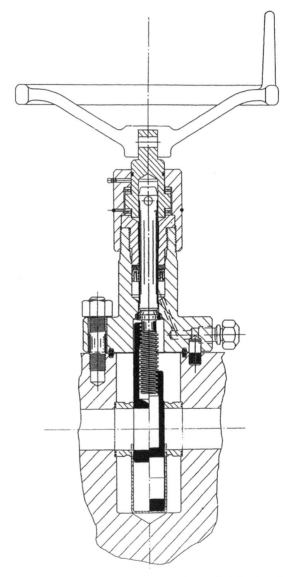

Fig. 3.12   A non-rising stem (internal screw) slab gate valve
(*Courtesy of Cooper Oil Tool*)

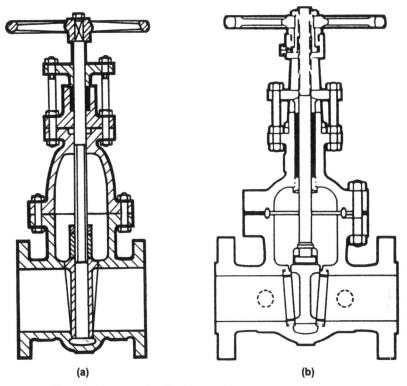

(a)   (b)

Fig. 3.13a   Wedge gate valve (inside screw)
(*Courtesy of British Valve and Actuator Manufacturers' Association Ltd*)

Fig. 3.13b   Wedge gate valve (outside screw), solid wedge
(*Courtesy of Hattersley Heaton Ltd*)

may be specified. The latter has a two-piece gate which can adjust to changes in seat angle whilst maintaining a good seal. A parallel slide valve (see below) will, however, usually be a better choice for such applications, except where a high degree of seat leak tightness at low pressures is a priority.

Wedge gate valves may have seating problems on dirty service due to material collecting on seats or in the base of the valve, but they usually give a better life in such service than soft seated ball valves. Services with

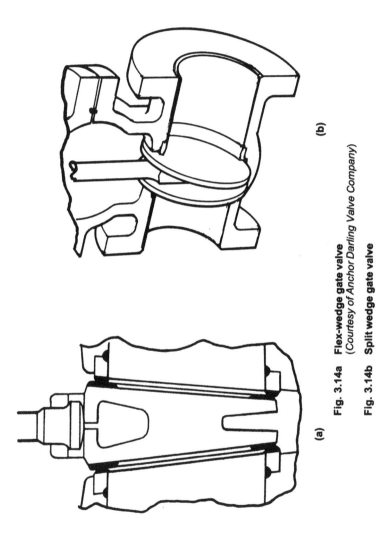

(a)

(b)

Fig. 3.14a   Flex-wedge gate valve
(Courtesy of Anchor Darling Valve Company)

Fig. 3.14b   Split wedge gate valve

abrasive particles or applications where wire drawing is possible will require hard faced wedges and seats. Conduit or parallel gate types give increased service life when used with fluids containing solid particles, because the gate cleans the seat when sliding over it and there is less chance of solids entering body cavities.

Some special rubber-sealed designs have good sealing characteristics when used on applications containing solids but have limited pressure and temperature range; other soft seats may be damaged by hard particles.

Flat sided or oval bodied designs (Fig. 3.15) are economical in terms of space and cost but their use should be restricted to the low pressure ratings.

Fig. 3.15 Oval bodied gate valve
(Courtesy of Stockham Triangle)

3.4.2　Parallel double disk gate valve (expanding wedge)

This valve (Fig. 3.16) has parallel seats and a split gate with an internal wedging action which forces the two gate halves out against the seats at

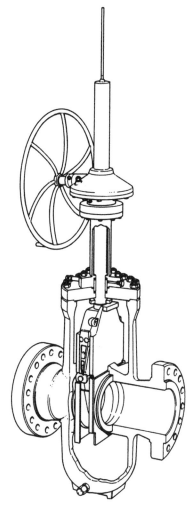

Fig. 3.16　Parallel double disk (expanding wedge) gate valve
(*Courtesy of Cooper Oil Tool*)

point of closure, thus providing a tight seal for liquid or gas service without the assistance of fluid pressure (Fig. 3.17). Through conduit versions are available. Because of this feature, the valve is suitable for single valve, double block and bleed duty when provided with a cavity bleed.

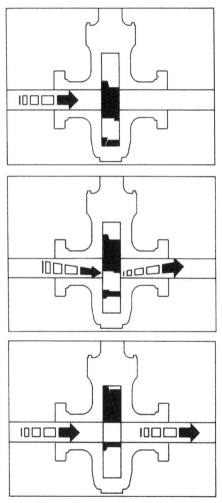

Fig. 3.17   Method of operation of parallel double disk (expanding wedge) gate valve
(*Courtesy of Cooper Oil Tool*)

This valve type should not be used on steam service, as the increased differential pressure resulting from condensate forming between the disks may result in leakage.

3.4.3   Parallel slab/through conduit gate valves

This valve has a single parallel faced slab gate, incorporating an aperture the same diameter as the valve bore (Fig. 3.18). When the gate rises to the fully open position it allows free and uninterrupted flow. The body cavity is sealed off by the gate when the valve is either fully open or fully closed. Valves may have rising or non-rising stem, can be obtained with a reduced

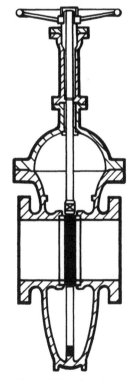

Fig. 3.18   Slab type through conduit gate valve
    (Courtesy of British Valve and Actuator Manufacturers' Association Ltd)

bore, and are available in reverse acting versions where the gate rises to close the valve (utilizing the unbalanced pressure force on the stem).

The seats are usually pressure-energized onto the gate (cf trunnion mounted ball valve) and double block and bleed designs are available which utilize a floating gate to seal both upstream and downstream seats simultaneously. Metal seated versions of the valve are amongst the most suitable for dirty or abrasive service.

The basic design is suitable for use on a wide range of applications, e.g., well head isolation (Fig. 3.19), large diameter piping, storage tanks, and on pipeline service where pigs may be passed through the line (Fig. 3.20).

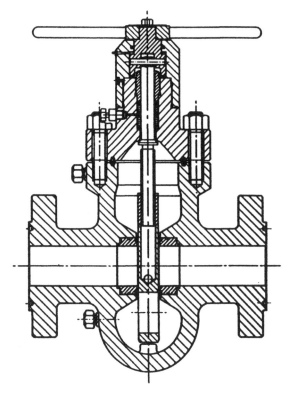

Fig. 3.19 Through conduit slab gate valve for well-head isolation duty
(*Courtesy of BEL Valves*)

Fig. 3.20   Through conduit slab gate valve for low pressure pipeline isolation
(*Courtesy of Daniel Valve Company*)

3.4.4 Parallel slide valves

Parallel slide gate valves (Fig. 3.21) maintain a high degree of fluid-tightness without the aid of a wedging action. This is achieved by two sliding disks which are maintained in close contact with the seats by a non-corrodable spring when not under pressure. Effective closure is obtained by pressure of fluid forcing the downstream disk against the mating body seat. Because of this, the valve will not provide tight shut-off at very low or zero differential pressure.

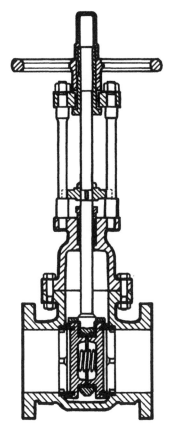

Fig. 3.21 Parallel slide gate valve
(*Courtesy of British Valve and Actuator Manufacturers' Association Ltd*)

On opening, the disks slide over the seat faces completely clear of the bore, thus permitting full flow through the valve.

These valves are recommended for applications where good shut-off characteristics are required for steam headers, steam isolation, feed water, and blowdown applications (where they are used in tandem with a sacrificial globe valve).

It should be noted that parallel slide valves are closed and opened according to stem position, not torque as with the wedge gate design. Because of this, and because seat faces are relatively wide, sealing capability and operating forces are virtually unaffected by wide temperature variations.

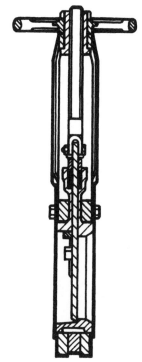

Fig. 3.22   Knife edge gate valve
  (*Courtesy of British Valve and Actuator Manufacturers' Association Ltd*)

3.4.5   Knife-edge gate valve

This valve has a bevel or knife-edged gate (Fig. 3.22, opposite), and is designed to handle slurries etc., liable to obstruct a wedge gate. The knife-edge pushes aside or cuts through solids in the flow.

The valve is usually designed to manufacturer's standard and is generally only suitable for low pressure applications.

When provided with a gas tight seal these valves are sometimes used on flare systems.

3.4.6   Compact steel gate valve (extended body)

These are small (<DN50) valves having one end extended to permit direct threaded or welded attachment to the pipe and a threaded or socket weld female connection on the other end (Fig. 3.23).

The valve may be used as a primary isolator for pressure measurement, vents, or drains. Use of the valve eliminates a pipe nipple and welded joint where applicable.

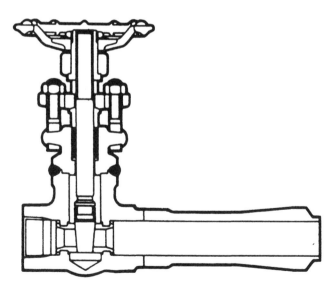

Fig. 3.23   Compact gate valve with extended end
(*Courtesy of Hattersley Heaton Ltd*)

3.4.7 Venturi gate valve

In this design (Fig. 3.24) the seat openings are smaller than the end ports and there is a smooth, conical transition between the two. The pressure drop through the valve can be up to twice that of a full bore valve but is generally negligible in relation to the whole piping system.

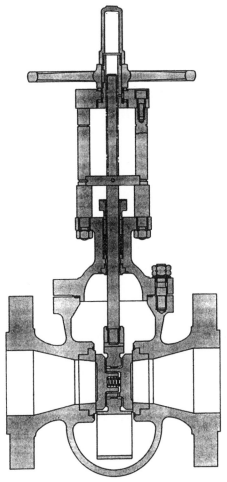

Fig. 3.24 Venturi-type gate valve (parallel slide)
(*Courtesy of Weir Hopkinsons Ltd*)

The advantages of this valve compared to the full bore type are: lighter weight; lower cost; and lower operating forces. Valves are usually of the wedge gate or parallel slide type.

When valves are installed in horizontal pipe runs line drainage may be necessary.

3.5 GLOBE OR SCREW-DOWN STOP VALVES

The globe or screw-down stop valve is used for flow regulation or as a block valve where high resistance to flow is not a disadvantage and a positive closing action is required. The valve has a tortuous flow path which is typical of the closing method and results in a higher resistance to flow than most other valves (Fig. 3.25). The configuration of the flow path is normally only suitable for uni-directional flow. High unbalance forces on single seated disk or plug designs tend to prevent opening with reverse flow but may be reduced in double seated designs.

Three body patterns are available:

(a)   standard (straight) (Fig. 3.25);
(b)   oblique (Fig. 3.26);
(c)   angle (Fig. 3.27).

All the above body patterns can be provided in needle valve form (Fig. 3.28). The oblique- and angle-type have much lower flow resistance than the straight-through globe.

Standard, oblique, or angle valves may be used to advantage for frequent on–off operation on gas or steam service because of the relatively short disk travel. There are two commonly available seat designs.

(a)   The first type has a flat seat and disk sealing metal to metal or using a soft seal ring incorporated in the disk or seat. A soft seal ring type may be specified where foreign matter might prevent tight closure or score sealing surfaces of metal-to-metal seats. The soft seal may be easily replaced. Soft seal rings are limited as to maximum allowable temperature.

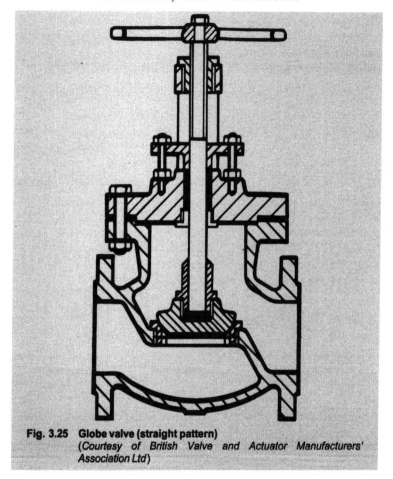

Fig. 3.25　Globe valve (straight pattern)
(Courtesy of British Valve and Actuator Manufacturers' Association Ltd)

(b)　The second type has a disk with a tapered or spherical seating surface, which provides a narrow line of contact against a conical seat. The narrow contact area tends to break down hard deposits that may form on the seat. This design is more suited to the higher pressure ratings.

Oblique-type valves have a relatively straight flow path, and are suitable for on–off or throttling duty on abrasive slurry or highly viscous

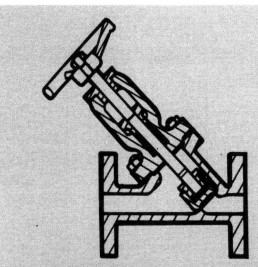

Fig. 3.26 Globe valve (oblique pattern)
(Courtesy of British Valve and Actuator Manufacturers'
Association Ltd)

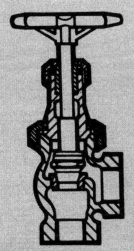

Fig. 3.27 Globe valve (angle pattern)
(Courtesy of British Valve and Actuator Manufacturers'
Association Ltd)

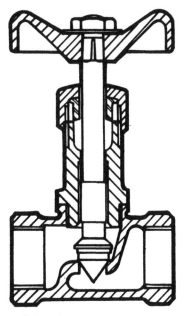

Fig. 3.28   Needle-type globe valve
(*Courtesy of British Valve and Actuator Manufacturers' Association Ltd*)

liquids. Standard-type valves are not recommended for these services. Flow resistance of oblique pattern valves is only about 25 percent of that of comparable standard-type valves.

Angle-type valves, when fitted at a change in direction of piping, save one bend or elbow and have the advantage of a smaller pressure drop than that for a standard pattern globe valve. These valves are not, however, extensively used because:

– the 90 degree bend in process piping may subject the valve to considerable stress at operating temperature;
– the handwheel may be placed in only one position with respect to the piping.

A further variation is the piston type globe valve where a radially sealed piston is used to cover and uncover ports in a cylinder or cage mounted within the valve body. Applications are governed largely by the material used for the piston seals (Fig. 3.29).

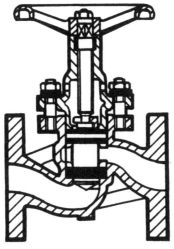

Fig. 3.29   Piston-type globe valve
(*Courtesy of British Valve and Actuator Manufacturers' Association Ltd*)

3.6   PLUG VALVES

Plug valves have quarter turn operation. Plugs may be tapered (Fig. 3.30) or parallel (Fig. 3.31) and in standard form are suitable for most on–off non-abrasive process and utility services and some slurries. They have straight through configurations typical of the sliding method. When used for throttling, special trim is necessary. Only full bore, round port valves are suitable for pigging and when valves are required for this duty the manufacturer should be consulted. Most designs have a reduced area of flow through the plug. Lubricated types require regular maintenance and PTFE sleeved and lined designs have temperature limitations.

Ideally, all types of plug valve should be regularly exercised to prevent operating torque becoming excessive.

Certain designs of fully lined or sleeved plug valves have excellent leakage performance, both down the line and to atmosphere, and are specifically used for caustic, chlorine, scaling, and similar services. The method of fitting or keying in the PTFE sleeve of sleeved plug valves is important if creep and high operating torque are to be avoided.

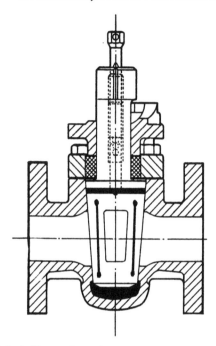

Fig. 3.30   Lubricated taper plug valve
(Courtesy of British Valve and Actuator Manufacturers' Association Ltd)

Lubricated types have been traditionally used for caustic, town gas, and sometimes for maintenance-compressed air services. Valves for chlorine service must have a drilling to vent the plug and any body cavity to the upstream port.

Fire tested glands can be obtained for sleeved valves to prevent leakage to atmosphere during or after a fire. Sleeved plug valves cannot, however, seal down the line once the sleeve is damaged.

On dirty service the seats will normally be wiped clean. There are usually no cavities for trapping solids and the sleeve can sustain some damage before, eventually, leakage occurs. However, dirty service could increase the already high operating torque.

Plug valves are made in four patterns, with port shapes and areas as follows (Fig. 3.32).

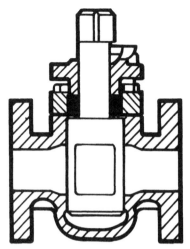

Fig. 3.31   Parallel plug valve
(Courtesy of British Valve and Actuator Manufacturers' Association Ltd)

(*a*)   *Round opening pattern*
   This has full bore round ports in both body and plug.
(*b*)   *Regular pattern*
   This has seat ports of rectangular or similar shape which have a
   cross-sectional area approximately 75 percent of the pipe.
(*c*)   *Venturi pattern*
   This has reduced area seat ports of round, rectangular, or similar
   shape. It is less expensive, with lower operating torque requirements
   than a regular pattern valve but with increased flow resistance.
   Recommended for use in larger sizes.
(*d*)   *Short pattern*
   This has reduced area seat ports of rectangular or similar shape, with
   a face-to-face dimension corresponding to that of wedge gate valves.
   Not recommended in larger sizes because the short length results in a
   small port area with abrupt change of throat shape between the
   flanges and plug.

   Valve plugs may be lubricated or non-lubricated. In the former type
lubricant is injected under pressure between the plug face and body seat to

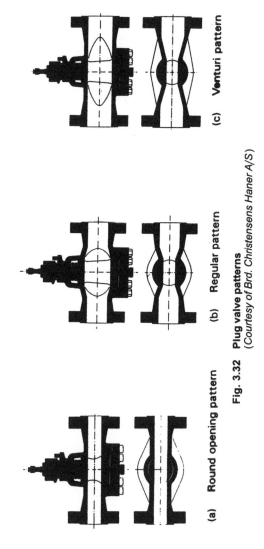

(a) Round opening pattern   (b) Regular pattern   (c) Venturi pattern

Fig. 3.32 Plug valve patterns
(Courtesy of Brd. Christensens Haner A/S)

reduce friction, provide port sealing, and to permit sealant jacking action to unseat the plug in tapered designs.

Non-lubricated plug valves incorporate mechanical design features to reduce friction between the plug face and body seat during operation by lifting the plug, or, in the split plug type valve, by withdrawing the plug halves from the seat prior to rotation. Alternatively, PTFE linings or sleeves are used. Both types are available in parallel and taper versions.

Plug surfaces may be:

- hard faced or hardened by heat treatment to prevent galling;
- plated;
- provided with an elastomeric coating;
- provided with soft seat inserts.

Sleeved plug valves (Fig. 3.33) are usually of the tapered design and incorporate a polymeric sleeve (usually PTFE) in the body. Normally only available in the lower pressure ratings, they can provide good leak tightness.

Fig. 3.33   Sleeved plug valve
(*Courtesy of Xomox Ltd*)

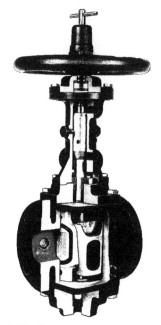

Fig. 3.34　Expanding plug valve
(*Courtesy of Orbit Valve plc*)

Expanding plug valves (Fig. 3.34) have a parallel design and incorporate a split plug with an internal wedge mechanism which is used to force the plug halves (slips) against the seats on closure and to release them on opening. Soft seal rings are usually employed and the design is capable of excellent seat tightness. These valves are suitable for use in double block and bleed applications.

Lifting (or wedge) plug valves (Fig. 3.35) are of the tapered design and utilize an operating mechanism whereby the plug is lifted from the seat prior to turning open or closed through 90 degrees, the object being to reduce operating torque whilst maintaining good sealing capability.

Eccentric plug valves (Fig. 3.36) are of the parallel design and utilize a cam action to drive a half-plug onto the downstream seat (Fig. 3.37). Designs are limited to the lower pressure ratings and lined versions are available.

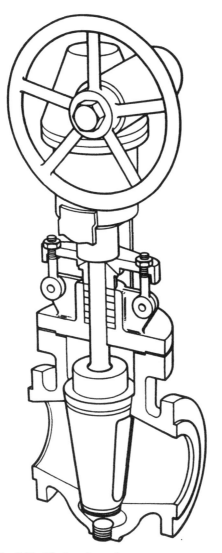

Fig. 3.35 Wedge plug valve
       (*Courtesy of Goodwin International Ltd*)

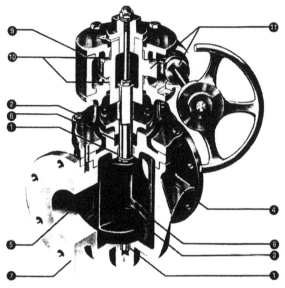

Fig. 3.36　Eccentric plug valve. 1 Bearing; 2 Bonnet bolts; 3 Plug; 4 The body; 5 Inlet port; 6 Gland packing; 7 Inlet flange; 8 Seat; 9 Gear box; 10 Gear box bearings; 11 Worm and wheel
(*Courtesy of de Zurik*)

Lined plug valves follow the tapered design and are fully lined (plug and body) for chemical resistance (Fig. 3.38).

Parallel plug valves which rely on lubrication to seal and protect the seats may be subject to through leakage. For bubble tight service, parallel plug valves should be of a soft seated design rather than a lubricated type.

The following limitations of lubricated plug valves should be noted:

(a)　they require a lubrication programme to be maintained;
(b)　the lubricant must be selected to suit the fluid;
(c)　the valve is subject to the temperature limitations of the lubricant;
(d)　the valve is unsuitable for throttling duty;

(e) lubricant may be washed from the plug face to contaminate the process stream;

(f) some process fluids of very low lubricity may dissolve the lubricant from the plug so that the valve may tend to gall;

(g) lubricant may entrap abrasive particulate in the working fluid.

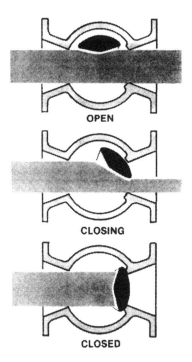

OPEN

CLOSING

CLOSED

Fig. 3.37   Operation of eccentric plug valve
(*Courtesy of de Zurik*)

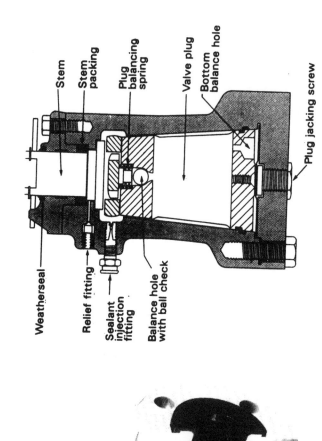

Stem

Stem packing

Plug balancing spring

Valve plug

Bottom balance hole

Plug jacking screw

Weatherseal

Relief fitting

Sealant injection fitting

Balance hole with ball check

(b)

(a)

Fig. 3.38a   Lined plug valve
(Courtesy of Xomox Ltd)

Fig. 3.38b   Balanced plug valve

Tapered plug valves have the following limitations:

(a) they may be difficult to free after prolonged setting in one position;

(b) conventional designs are unsuitable for most dirty or abrasive service applications unless special surface treatments are applied;

(c) simple designs have inherently high operating torque but this may be reduced by using a lift type valve, a low friction lining, a thrust bearing or, more commonly, by balancing the pressure across the plug (Fig. 3.38(a));

(d) sleeved or lined valves are subject to the temperature limitation of the elastomer/polymer sleeve or plug coating.

Soft seated valves should be fitted with an anti-static device.

Soft seated lift type plug valves may be considered for tight shut-off applications. For abrasive service a metal-to-metal seat lift type plug valve is available, but may require seat flushing. Valves of this type used on coking duty are flushed with steam.

When used on liquid service, manually operated valves located such that rapid closure could produce water hammer should be operated through a reduction gear.

Multiple port plug valves are available for diverter service. These may simplify piping layout, reduce the number of valves required, and eliminate elbows or tees.

The multiple port arrangement may be such that one channel closes before another begins to open preventing mixture of fluids or loss of pressure. Alternatively, some valves have greater port width so that in turning the plug, a new channel begins to open before the former channel is completely closed. This alternative may be used where it is necessary to carry out switching operation without stopping the flow at any time.

Multiple port valves can only provide closure when the operating pressure holds the plug against the body port which has to be shut off from the higher pressure or when the valve is of the expanding plug design. Leakage may occur when the operating pressure tends to push the plug away from the body port which has to be shut off. For vacuum applications ensure that the vacuum tends to hold the plug against the body port which has to be shut-off.

Four-way valves are intended for directional control only, and cannot be expected to hold high differential pressure without leakage from one side of the valve to the other.

For corrosive services lined valves are usually specified.

3.7   DIAPHRAGM VALVES

Diaphragm valves are used for block and control functions. The closure is a resilient diaphragm seating in the valve body. This diaphragm also provides the joint gasket between the body and bonnet and often the stem seal as well. Diaphragm valves are either manually operated by a handwheel closing device or by fluid pressure, normally air. Manually operated valves are typically of multi-turn operation.

Diaphragm valves have straight through flow or weir configurations and may be used for low pressure isolation or regulation of most gases and liquids, e.g., slurries, viscous fluids, and fluids which are chemically aggressive. They are supplied with various types of diaphragms and linings. Standard valves are normally supplied with malleable cast iron bodies which are not generally acceptable for petrochemical duties, but steel valves are available. Diaphragms are subject to wear and frequent maintenance may be required for regularly-used valves. Operating pressures and temperatures are severely restricted by the limitations of the membrane.

Diaphragm valve body configurations:

– weir types;
– straight through types.

The following are the standard body configurations.

(a) *Weir type valve (Fig. 3.39):*
     (i)   provides tight shut-off with comparatively low operating force
           and short diaphragm movement;

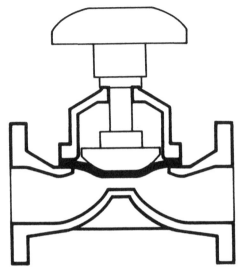

Fig. 3.39 Weir-type diaphragm valve
(Courtesy of British Valve and Actuator Manufacturers' Association Ltd)

    (ii)   has longer diaphragm life than straight through type, with reduced maintenance;

    (iii)   is better at throttling flow than straight-through type, although flow control is poor at very low flow rates.

(*b*)   *Straight-through type valve (Fig. 3.40):*

    (i)   is better than weir type when handling viscous fluids, thick slurries, and fluids containing deposits;

    (ii)   has longer diaphragm movement, which decreases diaphragm life, increases maintenance and, (because a more flexible diaphragm is required) limits the material choice to elastomers.

Valves have operating temperatures limited by the material used for the diaphragm or body lining. Operating temperature is normally between −65°C and 200°C.

    Diaphragm valves generally do not require gland packing and, having only three main parts – body, diaphragm and bonnet assembly, may be quickly dismantled for maintenance. Diaphragms may be changed in situ.

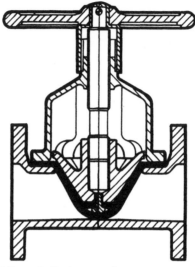

Fig. 3.40 Straight-through diaphragm valve
(*Courtesy of British Valve and Actuator Manufacturers' Association Ltd*)

For corrosive or toxic service a special bonnet, with a secondary stem seal to prevent leakage in the event of diaphragm failure, should be specified.

3.8 PINCH VALVES

Pinch valves have straight through configurations and are basically a reinforced rubber or elastomer tube or sleeve in a housing which is pinched for closure (Fig. 3.41).

The sleeve may be exposed with flanged ends but generally it is encased in a metal body. It is suitable for both pneumatic or hydraulic control. An enclosed valve may also be used for vacuum service subject to the manufacturer's approval.

Pinch valves are suitable for fine control or on–off operation on abrasive slurries, fluids with suspended particles, powders, or corrosive chemicals.

Valves are easily maintained with occasional replacement of the sleeve.

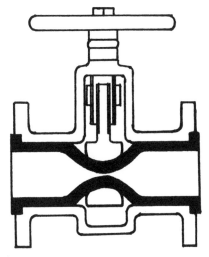

Fig. 3.41   Pinch valve
(Courtesy of British Valve and Actuator Manufacturers' Association Ltd)

Pressure and temperature ratings are restricted by the sleeve material. Valves are supplied to manufacturer's standards.

3.9   SAMPLING VALVES

Sampling valves are small bore valves generally similar to flush bottom valves, designed to draw off a small sample of fluid from process streams (Fig. 3.42). They are usually screwed into a half coupling or threadolet-type fitting welded to a vessel or pipe. Valves are suitable for use with liquids or slurries, and because they are flush bottomed they create a minimum of turbulence.

The valve design assures a free flowing sample, because the piston takes up the whole interior of the valve in the closed position so that sediment cannot accumulate.

The piston travels through a PTFE seal which may allow leakage unless the gland packing is compressed correctly. With too much compression the seal will flow inward, preventing reinsertion of the piston. As operation of the valve requires long piston travel, it will be slow to open or close.

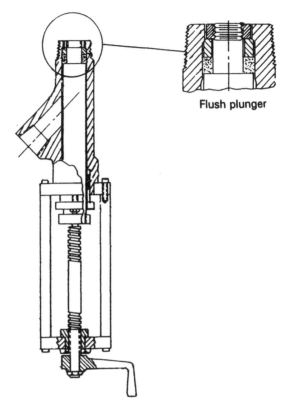

Flush plunger

Fig. 3.42 Sampling valve
(*Courtesy of Fetterolf Corporation*)

This type of sampling valve is mainly used when a conventional valving arrangement is not suitable, e.g., where plugging of flow passages can occur. Other types of sampling valve are available, generally as small bore instrument manifold blocks.

3.10 SLAM-SHUT VALVES (Fig. 3.43(a))

A valve in which the closure member (usually a swing disk) is held open against the flowing fluid. When released, it moves (owing to a combination of gravity and the dynamic effect of the flowing fluid) to close off against the seat. These valves are used as safety shut-off valves in gas systems and, unlike the excess flow valve, rely on the provision of an external signal to initiate closure. They are usually held open by a lever actuated by air pressure or a solenoid and are generally arranged so that loss of the control signal results in closure of the valve ('fail shut'). Wherever slam-shut valves are used, careful consideration should be given to the consequences of spurious closure.

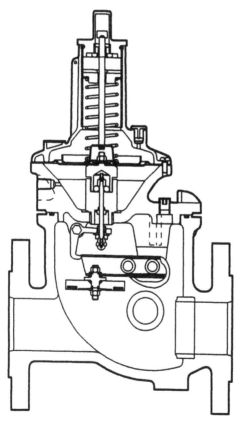

Fig. 3.43a   Slam shut valve
(*Courtesy of Bryan Donkin*)

3.11   ROTATING DISK VALVES (Fig. 3.43(b))

This is a variation on a parallel slide valve in which the (hard faced) disk is moved across the (hard faced) seat by rotary action of an arm attached to the valve shaft. The basic design employs a single, downstream seated disk and provides good sealing in abrasive or slurry service. Double disk designs are available offering sealing in both directions and there is

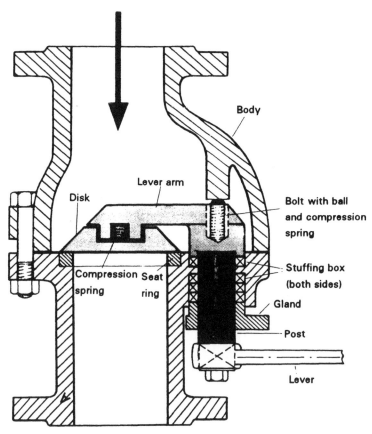

Fig. 3.43b   Rotating disk valve
(*Courtesy of Everlasting Valve Company Incorporated*)

another alternative in which the disk is wedged into the closed position by a lug located in the body.

3.12 DIVERTER VALVES

(Other terms in use include multi-port or switching valve; also change over valve.)

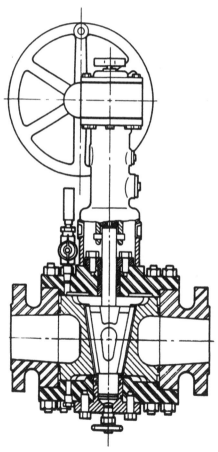

Fig. 3.43c Section through four-way diverter expanding plug valve
(*Courtesy of General Valve Company*)

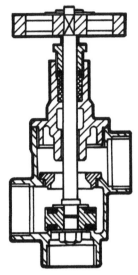

Fig. 3.44   Diverter type globe valve (three way)
(*Courtesy of British Valve and Actuator Manufacturers' Association Ltd*)

The principal function is to divert one or more flow streams whilst preventing intermixing (Fig. 3.44).

Diverter valves can replace several block valves but are not commonly used except in chemical process applications. Multi-port designs may vary from three to five ports according to requirements and are normally restricted to the lower pressure ratings (Fig. 3.43(c)). Operation is normally by manual intervention either directly or indirectly, e.g., powered actuators.

The types of valve that are suitable for diverting flow are limited to:

– plug valves (most common);
– ball valves;
– globe valves.

3.13   CCU SLIDE VALVES (Fig. 3.45(a))

These are large, low pressure gate valves usually of the slab type (although double disk versions are available) which are used for the control and isolation of hot, abrasive, gas and catalyst in the catalytic cracker units of oil refineries. Temperatures are extremely high (typically 700°C) and the valves are usually refractory lined and/or water cooled. Steam purging facilities are frequently provided. Design pressures are low and the valves are generally of fabricated construction. Three way diverter valves functioning on the same principle as the rotating disk valve (above) are also available.

3.14 BODY/BONNET/COVER JOINTS

The joint between the body and bonnet of a valve (or between the body and end connector, in the case of some ball valves) usually takes the form of a bolted closure of some kind. Ball valves often utilize a self-energizing gasket such as an elastomer 'O' ring or reinforced polymer lip seal. Such arrangements require only a modest amount of force to compress the seal so most of the bolt pre-load is available to prevent the joint opening under pressure. These types of seal exhibit high integrity under all pressure conditions, but are limited to applications where temperatures are relatively modest. For this reason they are also used in other valve designs where the incorporation of plastic or rubber parts already imposes a temperature limitation.

Gate, globe, and check valves generally employ a compressed gasket to seal the body bonnet joint and this is usually made from a material (e.g., steel, graphite, asbestos) which will withstand the maximum temperature rating of the valve. Full face sheet gaskets are used at the lowest pressure ratings whilst controlled compression, spiral wound gaskets are favoured for higher pressures. Iron or steel ring-type joints are used (especially in the petroleum industry) at the highest pressure ratings.

Where pressures are high body/bonnet flanges and bolting can become massive. The 'pressure seal' joint (Fig. 3.45(b)) uses a 'floating' bonnet with a gasket and segmental ring to react the upwards pressure force in the valve body wall, thus obviating the need for large flanges and bolts. The gasket in such joints used to be a plated iron or steel ring which was often difficult to remove after being in service and tended to leak at low pressures. Modern designs use a compressed graphite gasket which is easy to remove and has superior low pressure sealing ability.

(a)   (b)

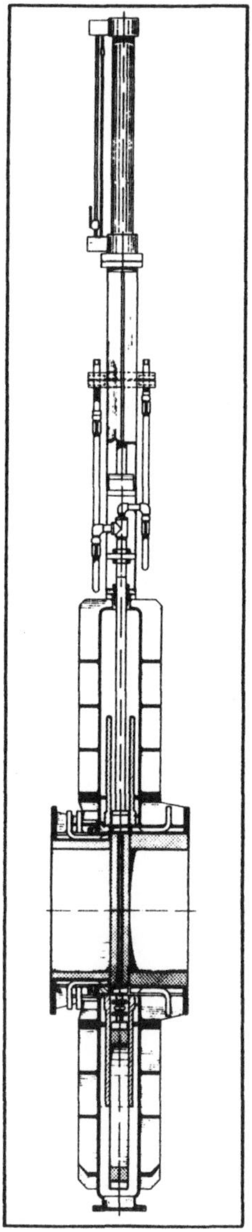

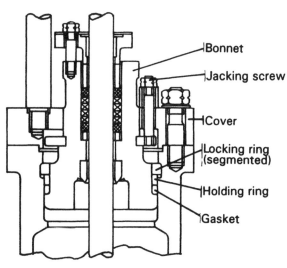

Bonnet

Jacking screw

Cover

Locking ring
(segmented)

Holding ring

Gasket

Fig. 3.45a CCU slide valve
(*Courtesy of Zimmerman and Jansen GmbH*)

Fig. 3.45b Pressure seal bonnet
(*Courtesy of Weir-Hopkinsons Ltd*)

CHAPTER 4

Valve Operation and Isolation

4.1 OPERABILITY

Manual valves are closed by turning a handwheel, wrench, or geared handwheel in a clockwise direction. Wrench-operated valves should be provided with limit stops to prevent overtravel.

Handwheels and wrenches should be constructed of steel or malleable or nodular iron. Pressed or stamped handwheels and wrenches are not recommended.

Butterfly valves with wrench operators should be designed to hold the disk in at least five equally-spaced intermediate positions between closed and fully open.

Handwheel keys should only be used in emergencies or where handwheels are known to be inadequately sized; they should be used with caution. Increased operating torque is usually indicative of a problem which needs to be addressed.

The effort required to operate a valve will depend upon its design, operating conditions, and size. When the effort to operate the handwheel exceeds 350N, geared operators should be considered.

All gear operators should be totally enclosed and suitable for the site conditions (e.g., onshore, offshore, and subsea). Any lubricants used should be suitable for use at the site ambient temperatures.

The following table provides guidance on valve sizes/pressure Class (Cl) ratings above which gear operators should be provided subject to the manufacturer's recommendations.

Valve type	Cl 150 (in)	Cl 300 (in)	Cl 600 (in)	Cl 900 (in)	Cl 1500 (in)	Cl 2500 (in)
Wedge gate	14	10	8	6	4	2
Globe	8	8	6	4	3	2
Ball	8	8	6	6	6	2
Butterfly	8	8	–	–	–	–
Plug	6	6	6	4	4	2

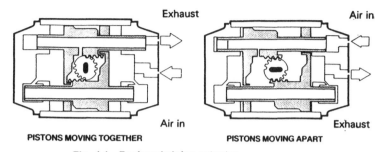

Fig. 4.1 Rack and pinion actuator
(Courtesy of Worcester Controls (UK) Ltd)

Parallel slide valves for steam services are normally provided with integral by-pass connection in size 8 NPS and above to equalize pressure on the disk before opening. The requirement for a by-pass is subject to the operating pressure and the manufacturer's recommendations. The by-pass pipe should meet the same specification as the associated piping specification.

Chainwheel operators should be used with caution and preferably avoided for valves in screwed lines, or for any valve smaller than 2 in.

In the case of actuated valves, the position of the valve closure member and direction of closure should be clearly indicated. Actuators should preferably be selected and provided by the valve manufacturer who can then take responsibility for the performance of the complete assembly.

When selecting actuators for quarter turn valves, the likelihood that the operating torque will rise during service should be borne in mind. Increases of two or three times the 'as new' value are not uncommon. Whilst the torque from rack and pinion designs (Fig. 4.1) is constant throughout the stroke, actuators employing a scotch yoke mechanism (Fig. 4.2) have a 'bath tub' torque characteristic and it may be that torque margins which are available at break-out are not maintained during rotation.

Electric actuators are available with remote or integral switchgear. The operating environment should always be considered in the latter case.

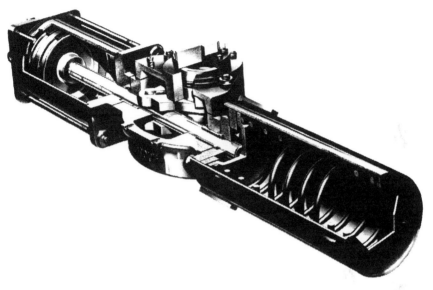

Fig. 4.2 Pneumatic actuator with spring return and scotch yoke mechanism
(*Courtesy of Rotork Controls Ltd*)

4.2 ISOLATION

4.2.1 Positive isolation
Where leakage cannot be tolerated for safety or contamination reasons positive isolation is made by blanking or the use of line blinds. It is important to realise that these operations are only possible if the isolating valve seals well enough to permit them to be carried out.

4.2.2 General isolation
Soft seated block valves, such as ball valves, sleeved plug valves, soft seated gate and butterfly valves can provide a good tight shut off for most clean services.

Metal seated valves will not usually achieve the same degree of seat tightness when new but will often maintain their performance more consistently. For high pressures/temperatures and dirty abrasive service they may be the only option and are the preferred type.

Fig. 4.3   Electric actuator with integral switch gear (shown separately)
(*Courtesy of Rotork Controls Ltd*)

Metal-to-metal taper plug valves have good isolation characteristics but usually require regular lubrication and maintenance.

For steam services parallel slide valves are recommended, especially at battery limits or as section isolating valves within any process unit where a section may be taken out of service for maintenance while the unit remains in operation. Because they rely on system pressure to make a seal, very low pressure sealing may not be effective. Wedge gate valves may be used as an alternative for general isolation duty or where good, low pressure leak tightness is required.

There are many variations on the solid wedge gate valve using split and self-aligning wedges for which various claims for improved sealing and wedge alignment are made. Solid wedges are normally specified for general service.

Butterfly valves vary greatly in design and are of varying degrees of leak tightness. Good results can be obtained from 'high performance' type using offset disks.

Double isolation of equipment is recommended for applications such as hydrogen service where leakage of a highly flammable searching medium is undesirable.

4.2.3 Block and bleed

This term describes the provision of a tapping point, either integral with the valve or located in the downstream pipework, which permits any leakage past a block valve to be bled off (Fig. 4.4(a)).

4.2.4 Double block and bleed

This term is used to describe the provision of two isolation points (either two separate block valves or a single block valve with two seats, each of

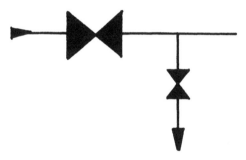

(a)   Block and bleed

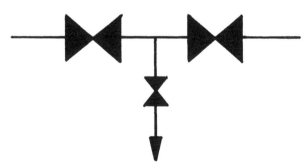

(b)   Double block and bleed

Fig. 4.4   Valve isolation arrangements

which makes a seal) with a tapping point located between them (Fig. 4.4(b)). The arrangement is used in two different ways:

(a) where the two valves or valve seats seal against a single source of flow or pressure (Any leakage from the first valve or valve seat is bled off through the tapping point, thus ensuring the integrity of the second seal.);

(b) where the two valves or valve seats seal against two separate sources of flow or pressure which are applied from opposite directions (Any leakage from either source is bled off through the tapping point, so preventing contamination or mixing of the two sources.).

When a single valve is used for this duty it should ideally be of a type where the seat load is applied mechanically so that it is independent of variations in line pressure. Suitable valve types include the parallel double disk gate valve with expanding wedge, the expanding plug valve and high integrity versions of the wedge gate valve (e.g., soft seated). Such valves (when provided with appropriate tappings) are suitable for either of the two applications described above although arrangements for pressure relief of the valve cavity must be made where liquids subject to temperature increase are likely to be trapped (see below).

Both trunnion mounted ball valves (Fig. 3.3) and through conduit slab gate valves (Fig. 3.18) can be used in single valve arrangements where simultaneous sealing against both upstream and downstream pressure is required. It should be noted that these designs rely on the line pressure to make an effective seal, and the use of springs, etc. to provide supplementary mechanical loading at low pressure is not always effective.

Through conduit slab gate valves can also be used where a double seal against upstream pressure is required, and they have the advantage that the pressure load applied to the upstream seat is transmitted, through the floating gate, to the downstream seat. Trunnion mounted ball valves with double piston effect seats are suitable for double block applications but should not be used for single valve double block and bleed duty where any one of the above alternatives is possible.

Butterfly, globe, and conventional plug valves are not suitable for double block or double block and bleed duty where a single valve is required since they effectively have only a single seat.

Certain hazardous applications (e.g., hydrogen service) require that two separate valves are provided in double block or double block and bleed arrangements.

4.2.5 Cavity relief

Valves with both upstream and downstream seats which can seal when a differential pressure is applied between the valve body cavity and the pipe need to be provided with some means of relieving cavity overpressure. This may arise when liquid, trapped in the cavity when the valve is closed, is heated (e.g., by a fire). Under these circumstances the generated pressure may be many times the valve rated pressure.

Relief can be provided by drilled holes (e.g., through the valve seat on the upstream side), pressure equalizing pipes, or external relief valves (Fig. 4.5). Trunnion mounted ball valves and slab type gate valves can usually be shown to be self-relieving across the seat (Fig. 3.8).

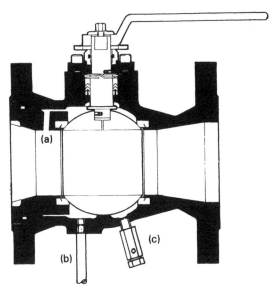

Fig. 4.5   Cavity overpressure relief of seat-supported ball valves; (a) drilling behind upstream seat; (b) connection to upstream pipework; (c) external relief valve
(*Courtesy of Hindle Cookburn Ltd*)

CHAPTER 5

Valve Types for Prevention of Flow Reversal (Check)

Check valves are required to permit forward flow and prevent reverse flow. This is achieved through linear or rotary (angular) motion of a closure member which is kept open by the flowing fluid. When the flow is reduced towards zero, or reversed, the closure member is moved against its seat by forces due to gravity, supplementary springs, and back pressure.

> Many types of valves are used. Installation and process design considerations are studied, followed by individual valve types including:
>
> - lift check – normally small disk, piston, and ball types;
> - swing check – normally DN50 and above;
> - diaphragm check – utilizes flexible diaphragm;
> - screw-down stop and check – globe and swing types with provision for manual closure;
> - wafer check – a narrow valve style for installation between flanges;
> - spring-operated non-slam check valve – axial flow type for pulsating flow etc;
> - foot valves – pump suction valves.

A position indicator mechanism is not usual and may be incorporated only on swing check valves and dual disk wafer-type check valves. This necessitates a penetration of the pressure-retaining boundary.

Only specially designed swing-type check valves are suitable for use in lines which have to accommodate pigs.

5.2 INSTALLATION AND PROCESS DESIGN CONSIDERATIONS

The self-acting principle of check valves introduces dynamic response as well as static (e.g., flow configuration) considerations, and valve characteristics should be such as to provide the following properties.

(a) Free and unrestricted movement of the closure member. Under normal operating conditions this should ideally be kept fully open against its stop. Oversizing should be avoided since this can lead to partial opening which may result in chatter of the valve against its seat or disk stop and excessive wear of active components.

    Manufacturer's sizing data should be consulted when selecting check valves.

(b) Full closure under back pressure conditions with adequate seat loading to minimize through leakage. In some cases, particularly where pressures are very low, it may be necessary to consider supplementary loading. This can be achieved in some designs by additional weighting or stronger spring loading of the closure member.

(c) A closure response which ideally follows reducing flow in the pipe such that when forward flow ceases, the valve is fully closed. A valve which permits reverse flow to occur can cause reverse rotation of pumps or compressors and check valve slam which itself results in pressure surge and possible damage to the valve, piping system and ancillaries. In general, surge should be considered when normal fluid velocities exceed 4.5 m/sec for liquids or 27.5 m/sec for gases and whenever operating pressures are close to the design pressure of the pipe system. The following points should be noted.

(i) Comparison of predicted against required speed of valve closure, closing force necessary to prevent excessive back flow, surge pressure etc. may be qualitatively assessed or subject to detailed analysis if very critical. Calculations should be based on process data and manufacturer's valve data. In general terms, a valve that closes at a mean velocity equal to or less than the normal flow velocity should avoid excessive surge pressure.

    Lower closing velocities may be acceptable under certain conditions, e.g., on single pump systems for long pipelines where the terminal back pressure and maximum elevation are low. High closing velocities may be necessary in parallel pumping systems to prevent back flow into a failed pump. Supplementary loading can improve the response of some valves.

(ii) Generally, check valves with a short displacement and low inertia of the closure member, supplemented by spring loading, provide the most rapid response and lowest shock loading at closure. Smaller sizes of valves generally provide the fastest closing response.

(d) Stable operation to avoid rapid fluctuation in movement of the closure member. This may be due, for example, to large variations in operating pressure (or smaller variations with pulsing flow), which can lead to valve chatter, excessive wear, and poor reliability. Damping devices such as dashpots can be fitted to some designs to improve valve stability.

Note that damping at final closure may also be required to prevent shock loading by slamming of the valve onto its seat. This may be a requirement in systems where extremely rapid flow reversals can occur, e.g., with compressible fluids.

In addition it may be necessary to consider the following.

(e) The frequency of flow reversal occurrences. If numerous, this may have an adverse effect on wear and reliability of valve components in some designs.

(f) Flow resistance (pressure drop). Resistance to flow varies widely in different designs and also generally increases with smaller valve sizes. Low resistance is achieved in some designs by valve patterns in which the closure member and seats are inclined towards, rather than normal to, the flow path. Generally, designs which provide easiest pigging etc., are most susceptible to inducing pressure surges and flutter.

(g) Location of the valve bore in either the horizontal or vertical plane. A vertical location can adversely affect the response of some types of check valves, depending on the pattern, and should be avoided if possible. The direction of normal flow may also affect closing force requirements. Generally, flow in an upward direction is preferred and flow in a downward direction should be avoided. Swing check valves should never be used with downward flow, and lift check valves should not be fitted in vertical pipes. The manufacturer should always be advised of the intended valve location and, if vertical, the direction of flow.

Check valves are extremely sensitive to upstream piping features and elbows, valves, and so on located immediately upstream can have a disastrous effect on performance. Such features should never be nearer than three pipe diameters to the check valve.

5.3 VALVE CATEGORIES

Check valves may be broadly divided into thee main categories. These comprise:

- the lift check valve, based on linear motion of the closure member;
- the swing check valve, based on rotary (angular) motion of the closure member;
- the diaphragm check valve which operates by flexing of a membrane.

There are a number of variations in design of all three types of valve. Characteristics of each design vary considerably and should be considered when selecting valves for particular duties.

5.4 LIFT CHECK VALVES

A lift check valve is one in which the check mechanism incorporates a disk (Fig. 5.1), ball (Fig. 5.2) or piston and cylinder (Fig. 5.3) (which provides a damping effect during operation).

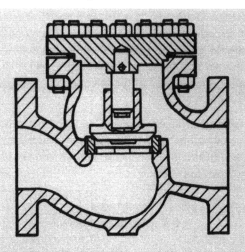

Fig. 5.1   Lift check valve (disk type)
(*Courtesy of British Valve and Actuator Manufacturers' Association Ltd*)

Fig. 5.2   Lift check valve (ball type)
(*Courtesy of British Valve and Actuator Manufacturers' Association Ltd*)

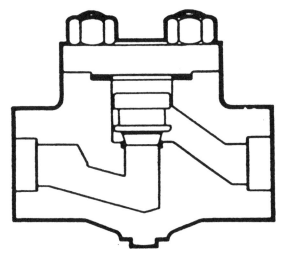

Fig. 5.3 Lift check valve (piston type)
(*Courtesy of Hattersley Heaton Ltd*)

Lift checks are mainly used on small bore piping systems (DN50 or less) where their relatively high flow resistance is acceptable. Larger disk check valves are sometimes used because of their improved response compared to swing checks.

Free and unrestricted movement may be difficult to achieve since the majority of designs depend on close guiding of the closure member. The ingress of dirty or viscous fluids can result in slow response or even jamming unless considered in the design, especially with piston and disk check valves. Ball check valves tend to be less affected due to freer guiding of the ball. For gritty liquids composition disks are available.

Full closure under back pressure conditions is often provided by supplementary spring loading due to the low mass of closure members in small lift check valves.

Closure response is potentially fast due to the inherent characteristics of a short travel (lift) compared with other types of check valve and the low inertia of the closure member due to its low mass. Thus lift check valves are suitable for many applications which could result in surge problems for other types of valves.

Stable operation avoiding rapid fluctuations in movement of the closure member cannot be provided by ball check valves which should not be used with widely varying pressures, pulsing flow, or frequent flow reversals. Piston and disk check valves may perform better but under severe conditions it may be necessary to consider a damping device.

Flow resistance is inherently high due to the tortuous flow path. However, where lower resistance is required patterns of 'Y' or oblique type are available. '

These valves should always be located in horizonal pipes with the axis of the disk, etc., in the vertical plane since they depend on the force of gravity acting on the closure member. Spring-loaded types may operate in any orientation, but manufacturers should be consulted.

Where the normal range of check valves is found to be unsuitable, e.g., in preventing excessive pressure surge or providing adequate stability with wide pressure variations, pulsing flow and frequent flow reversals consideration should be given to the use of axial flow check valves (see below).

5.5   SWING CHECK VALVES

A swing check valve is a check valve in which the mechanism incorporates a disk which swings freely on a hinge (Fig. 5.4).

The bearing assemblies for the hinge and disk are better shrouded from the flow stream so that dirty and viscous fluids are less able to obtain ingress and hinder rotation of the closure member.

Swing check valves are used over a much larger size range than lift check valves. Typically swing check valves are available from DN 50 up to DN 600 and greater, but weight and travel of the disk may become excessive in very large sizes and may require special designs for satisfactory operation. Rubber-lined disks are available for gritty fluids or where positive shut-off is required at low pressure.

Swing check valves may be specified for horizontal or vertical upward flow on low velocity or highly viscous fluids.

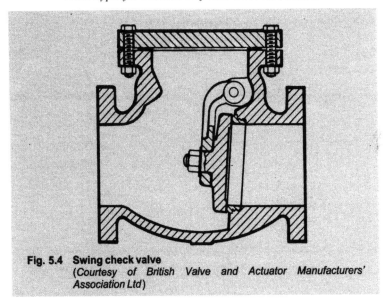

Fig. 5.4   Swing check valve
(Courtesy of British Valve and Actuator Manufacturers' Association Ltd)

These valves are unsuitable for frequent flow reversal or pulsating flow. When installed in a system sensitive to sudden flow reversal a balance weight or dashpot should be specified. A balance weight may also be required when the valve has to open under minimum pressure differential.

Full closure under back pressure conditions may be supplemented in the conventional and tilting disk designs by additional weighting of the disk or by an external weighted lever arm or spring attached to a shaft extension through the body (Fig. 5.5). The latter introduces the additional complication of a gland to seal the extension arm and may result in excessive closing force (shock loading) unless a damper is fitted.

The closure response of swing check valves is generally slower than with lift check valves. This is mainly due to the long travel of the disk during rotation from fully open to closed and to its inertia, depending on the mass and moment arm.

Conventional-type swing check valves in which the swing arm is hinged outside the flow path, have the slowest response. Although widely used, they are generally best suited to gravity flow or pumped (liquid) systems where flow velocities are relatively low and do not fluctuate. Improved

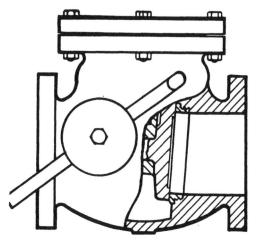

Fig. 5.5 Swing check valve with (external) supplementary loading (assisted closure)
(*Courtesy of British Valve and Actuator Manufacturers' Association Ltd*)

operating characteristics result from reducing the travel by inclining the seat and disk towards each other, and whenever possible the angle between the seat and the fully open position of the disk should be restricted to about 65 degrees. Another method of reducing inertia is by locating the hinge axis at the edge of the disk.

> Tilting disk check valves (Fig. 5.6) incorporate a disk which swings on a hinge which is parallel to, but just above the horizontal axis of the piping.

Tilting disk check valves are a variant of the conventional type and have a faster response by virtue of a shorter path of travel of the disk centre of gravity and reduced inertia. The flow resistance, however, is greater than that of conventional valves. Tilting disk check valves are generally used as 'non-slam' valves, since closure at instant of reversal of flow is almost attained.

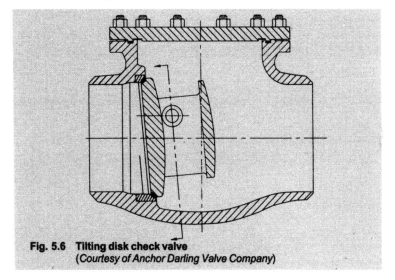

Fig. 5.6   Tilting disk check valve
(Courtesy of Anchor Darling Valve Company)

Split or 'duo-disk' valves (Fig. 5.7) also provide a fast response due to the short path of travel of the half disk centre of gravity and low inertia (resulting from the low mass, and short moment arm of the half disk allied to the use of closing springs).

Swing check valves of the split-disk-type depend on internal spring loading for closure, and supplementary loading can be provided by fitting stronger springs. They are usually made in 'wafer' pattern, and are better suited than the conventional type to the high flow velocities found in gas service.

Timing of closure cannot alleviate all shock conditions at a check valve. Typical cases are:

(a)   reduction of pressure at the valve inlet which causes the decelerating fluid to be flashed downstream from the valve;

(b)   stoppage of flow caused by sudden closure of a valve some distance downstream from the check valve, where this is followed by returning water hammer.

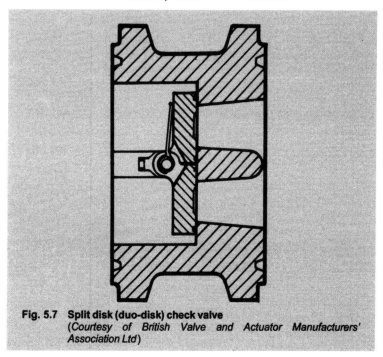

Fig. 5.7   Split disk (duo-disk) check valve
(Courtesy of British Valve and Actuator Manufacturers' Association Ltd)

For these applications slower closure may be necessary, in which case the check valve should be equipped with an internal dashpot.

Unstable operation with rapid fluctuations in movement of the closure member can be a problem with swing check valves. Where these conditions are unavoidable, stability of conventional and tilting disk types may be improved by fitting an external damping device (dashpot) via a shaft extending through the body (Fig. 5.8). Split disk check valves are more difficult to damp since external methods usually cannot be employed.

Generally, swing check valves should be avoided where unstable conditions are likely, e.g., wide velocity variations, pulsing flow, frequent flow reversals, and location close to pipe bends etc. Under certain con-

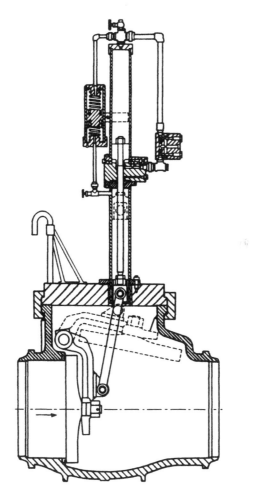

Fig. 5.8   Swing check valve fitted with dashpot
  (Courtesy of Anchor Darling Valve Company)

ditions fretting at the hinge axis and even mechanical failure may occur.

Low resistance to flow is a particular advantage of swing check valves due to the straight through flow configuration, subject to the angle of opening at operating conditions. Special designs of swing check valve may also be suitable for pigging. Split disk valves tend to have a proportionately greater resistance to flow as the size decreases, due to the body centre web (sealing each half of the split disks) taking up a greater proportion of the flow path.

Location in a vertical pipe requires careful consideration. Conventional and tilting disk valves must always be located with flow in an upward direction but the design must prevent the disk from reaching a stalled position when fully open or the valve will not close upon flow reversal. It should also be recognized that in the fully open vertical position the disk/swing arm has a very small closing moment, further reducing response (unless supplementary loading is used). Split disk check valves are more suited to vertical applications but preferably with flow in an upward direction. They may be considered for downward flow applications when stronger springs can be selected to suit the operating conditions. The manufacturer should always be advised of any such applications.

Sizing of check valves should ideally be such that, at minimum normal sustained flow, the check mechanism will be held against the stop in the fully open position. Applications in gas or steam lines, or in liquid lines with low or unsteady flow should be fully described in the purchase specification so that the manufacturer can evaluate the suitability of the valve design. The non-slam characteristics of check valves in compressor piping systems should be compatible with the compressor manufacturer's requirements.

Swing check valves may also be provided in the following additional types:

- screw-down stop and check;
- wafer;
- spring operated non-slam;
- foot.

5.6 DIAPHRAGM CHECK VALVES (Fig. 5.9)

Although less commonly used than conventional lift or swing check valves, diaphragm check valves are worthy of consideration when the pressure and temperature limitations of the flexible membrane which forms the closure member permit their use.

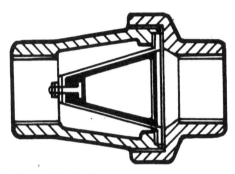

Fig. 5.9 Diaphragm check valve
(Courtesy of British Valve and Actuator Manufacturers' Association Ltd)

Free unrestricted movement, full closure, and fast closure response are all characteristics of the diaphragm check valve. It also has the ability to handle viscous or abrasive fluids and slurries more reliably than other types of check valve. Small sizes are usually of the cone type and larger sizes are typically of the nozzle type. The design can also provide stable operation with pressure variations, pulsing flow, and frequent flow reversals, but care is required in the selection of suitably durable materials for the membrane.

Location may be either with the bore horizontal or vertical. Data on flow resistance should be obtained from the manufacturer.

5.7 PISTON TYPE CHECK VALVES (Fig. 5.10)

The designs available include straight through and oblique patterns and some (usually larger size valves) where pilot pressure energizes the

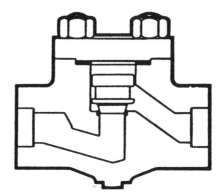

Fig. 5.10   Lift check valve (piston type)
(*Courtesy of Hattersley Heaton Ltd*)

piston to close during a reversal situation. Some general features are listed below:

(a)   generally used on line sizes below 2 in. NPS;
(b)   causes relatively high pressure drop;
(c)   spring loaded type may operate in any orientation, other types should be installed so that the piston will close by gravity;
(d)   suitable for frequent, but not sudden, flow reversal and pulsating flow duties;
(e)   unsuitable for gritty or dirty fluid service;
(f)   equipped with a dashpot and orifice to control the rate of movement of the piston (as the orifice required for liquid service is considerably larger than the orifice for a gas service changing services will involve a change of orifice).

5.8   SCREW-DOWN STOP AND CHECK VALVES (Fig. 5.11)

These are valves in which the disk may be held closed by a valve stem which can also be retracted to permit free movement of the disk. They are

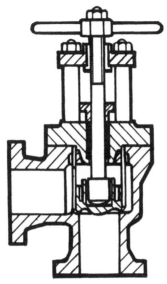

Fig. 5.11   Screw down stop and check valve
(*Courtesy of British Valve and Actuator Manufacturers' Association Ltd*)

generally used in steam generation by multiple boilers, where a valve is installed between each boiler and the main steam header. These are normally of the globe type, but swing types are also available.

5.9   WAFER CHECK VALVES (Fig. 5.12)

The general features of these valves are listed below.

(a)   These are swing-type check valves which are installed between pipe flanges and in which the mechanism consists of a single or dual plate swinging on a hinge.

(b)   The general characteristics are as noted for swing check valves.

(c)   Reduced bore types are generally available, in line sizes DN50 and above.

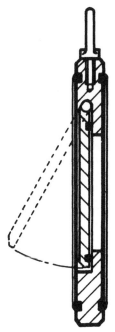

Fig. 5.12 Water check valve
(*Courtesy of British Valve and Actuator Manufacturers' Association Ltd*)

(d) Single plate types cause high pressure drop, and are unsuitable for use on low flow and low pressure gas services.

(e) These valves need to be removed from line for repair or inspection.

Wafer check valves should be used in flammable or toxic service with caution due to the possibility that, in the event of fire, expansion of the bolts may cause flange leakage. When a wafer check valve is used and a fire hazard exists, a light gauge steel shroud should be wrapped around the valve and exposed bolts, irrespective of service. Alternatively, select a lug-type valve in which the bolts pass through the body.

Care should be taken to see that the opening plates do not foul adjacent valves or connected pipework.

5.10 SPRING-ASSISTED, AXIAL FLOW, 'ANTI-SLAM' CHECK VALVES
(Fig. 5.13)

These are axial flow valves often used on compressor discharge lines subject to pulsating or low flow conditions where a tilting disk type valve may 'chatter'. They offer the least resistance to flow of any check valve type and their ability to follow variations in flow make them the least susceptible of all check valves to valve slam and pressure surge.

These valves are typically selected for onerous duties, providing a rapid closure by spring loading. Stability is claimed to be improved by use of the venturi principle in the design of flow passages through the body. Sliding parts are largely shrouded from the flowing process fluid by the central housing. These valves can be mounted either horizontally or vertically. The design is becoming more common and several manufacturers now produce it.

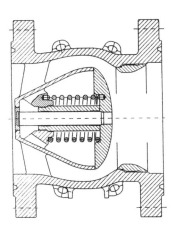

Fig. 5.13 Axial flow check valve
(*Courtesy of Mockveld Valves bv*)

5.11 PLATE CHECK VALVES (Fig. 5.14)

Plate check valves which use flexible metallic or polymer plates or membranes are normally fitted to compressors. This type of valve

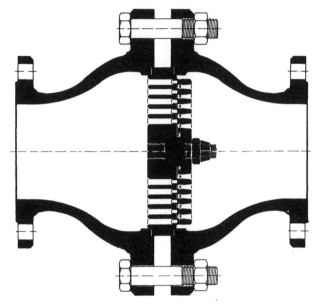

Fig. 5.14   Plate check valve
(*Courtesy of Hoerbiger Ventilwerke AG*)

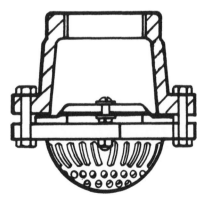

Fig. 5.15   Foot valve with strainer
(*Courtesy of British Valve and Actuator Manufacturers' Association Ltd*)

provides a very fast closing response and is particularly suited to pulsing flow with compressible fluids. The frequency of flow pulsations may require special consideration of design to avoid plate flutter. Generally, designs are limited to a low differential pressure across the valve. Valves are suitable for mounting either horizontally or vertically.

5.12 FOOT VALVES (Fig. 5.15)

These valves are generally installed at the suction inlet of a pump to maintain prime. The valve may be fitted with a strainer to keep dirt and other foreign matter from entering the pump suction.

CHAPTER 6

Valves for Special Applications

General considerations:

- safety;
- electrical isolation;
- special process applications;
- searching duties;
- materials of valves;
- special services.

6.1 GENERAL

- Soft seated valves (e.g., ball, plug, and butterfly valves) used in hazardous areas where they could be subjected to fire should be of a fire tested design. Metal seated valves may require a 'fire safe' gland (e.g., a graphite gland packing) and fire-resistant joint gaskets.
- For particularly hazardous service or high pressure, additional non-destructive examination may be specified for the pressure-retaining boundary. ANSI B16.34 provides guidance on the areas which should be considered and EEMUA publication 167 specifies requirements for three quality levels of steel valve castings.
- Electrical isolation of valve flanges may be necessary when mating with pipe flanges of dissimilar material. Similarly, electrical continuity and earthing may be required in fully-lined piping systems where static discharge can be a problem, particularly where gases are flowing at high velocities and liquid droplets or solid particles are present or where low conductivity liquids are involved.
- Oxygen service, cryogenic, and other special process applications may require valves to be thoroughly degreased, cleaned, and assembled in a 'clean room'.
- Valves required for searching duties (e.g., hydrogen) are often subjected to a helium leak test, or to a test using a mixture of air or nitrogen plus helium, to demonstrate a high degree of leak tightness.

- Materials of valves used in sour and/or chloride service should comply with NACE Std MR–01–75.
- Valves for services such as sulphur and applications where the fluid can solidify may need to be provided with steam jackets or trace heating.

6.2 LOW EMISSION SERVICE

This describes applications where escape of the working fluid to atmosphere is undesirable for environmental, safety, or economic reasons. The stem seal (gland) of rising stem valves needs particular consideration in such cases and specially-engineered arrangements of graphite and other materials are frequently employed, as is live (spring) loading, in an attempt to extend packing life and prevent relaxation occurring. Graphite packing should, however, be avoided if chemical oxidizers are present and should always be used in conjunction with a corrosion inhibitor where the valve stem is made from a martensitic grade of stainless steel (e.g., 410).

Mechanical aspects of the valve (such as straightness and surface finish of stems/stuffing boxes) must be controlled to a good standard, particularly where actuators are employed, and systematic plant maintenance routines are helpful in ensuring long-term performance.

In the case of smaller size valves (< DN150) a bellows seal may be used to prevent any escape of the working fluid to atmosphere (Fig. 6.1). The lower end of the bellows is sealed to the valve stem whilst the upper end is sealed to the bonnet. A secondary stem seal in the form of a packed gland or 'O' ring is often provided in case of bellows failure. Bellows seals are most often used in rising stem designs (gate, globe) but are also available for quarter turn valves (ball, butterfly, etc.).

6.3 CRYOGENIC SERVICE

Valves for cryogenic applications (below − 50°C) should comply with BS 6364 or equivalent standard and are normally provided with extended bonnets (Fig. 6.2).

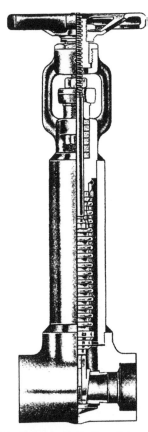

Fig. 6.1   Bellows seal gate valve
(Courtesy of Velan Incorporated)

Designs normally employed are gate, globe, ball, or butterfly types manufactured in stainless steel, monel, bronze, or cupro-nickel.

The extended bonnet allows a reasonable temperature gradient up to the gland and point of operation. It also allows a liquefied gas to reach a vaporizing temperature. The bonnet should provide means of venting any excess pressure build up, should this be necessary, and valves should normally be installed with stems vertical or at no more than 45 degrees to the vertical, to maintain a low conductivity vapour lock in the bonnet.

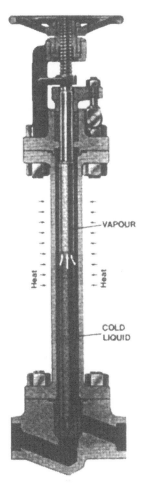

Fig. 6.2 Extended bonnet valve for cryogenic service
(*Courtesy of Velan Incorporated*)

Leakage rates normally acceptable in conventional application are not acceptable for cryogenic service where leaking fluid could freeze and affect the plant balance. Stem leakage may result in stem seizure and packing failure. Seats and seals are usually manufactured in KEL–F, PTFE, and similar materials and need careful selection for temperatures below −65°C.

Ball valves with soft seals or other valves having a closed body cavity may require facilities for relieving overpressure caused by thermal expansion of fluids as temperatures rise.

High energy shocks may occur in liquid oxygen systems: this causes debris to be dislodged from valve seats. Material should be chosen to eliminate fire risk, and stainless steel materials usually acceptable for oxygen service may not be suitable. Bronze or monel body and trim materials are recommended to prevent a spark occurring during high energy mechanical impact.

Cryogenic liquids are generally non-lubricating and, therefore, galling may occur between relatively soft metal mating parts; bronze stem bushing, KEL–F, PTFE or hard seats, special coatings, and solid film lubricants are all utilized to prevent this. For the same reason, operating torques of cryogenic valves are often higher than those of similar valves at higher temperatures.

All valves for cryogenic service should be cleaned to a high standard and be free of moisture and grease.

6.4   VACUUM SERVICE

For vacuum service soft seated valves may be specified, including high performance butterfly valves. For extremely high vacuum, metal-to-metal closing mechanisms are likely to be required.

Bellows stem seals may be specified, provided the cycle life is compatible with the application. Secondary packing should be specified in such cases.

For packed glands valve stems need to be truly round and parallel and have a smooth finish of 0.4 micrometres or even better. Similar attention needs to be given to the stuffing box bore.

Valve packing must be suitable for 25 mm Hg absolute pressure.

6.5 DELUGE VALVES

These valves are used on firewater deluge service. Proprietary deluge valves are preferred to process control valves for this service for the following reasons.

(a) They open virtually instantaneously, whereas a process control valve can require several seconds to operate.
(b) They fail safe despite damage to the pneumatic detection or actuation system. This may not be true in the case of process control valves.
(c) Process control valves are liable to seize when they stand inactive for extended periods of time as is common in deluge service; proprietary deluge valves are designed to avoid this problem.
(d) For any given line size, the deluge valve saves space and weight when compared with actuated process valves.

6.6 EXCESS FLOW VALVES (Fig. 6.3)

Excess flow valves are designed to close automatically when the flow through them exceeds a specified rate. They may be installed where fluid leakage through a ruptured line would cause serious damage.

These valves incorporate a spring-loaded valve disk which will only close when the forward flow of fluid through the valve generates sufficient force, or differential pressure, to overcome the power of the spring holding it open.

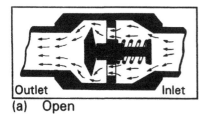
Outlet   Inlet
(a)   Open

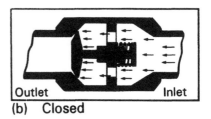
Outlet   Inlet
(b)   Closed

Fig. 6.3   Excess flow valve (a) open (b) closed
(*Courtesy of IPS Instrument Product Services*)

Each excess flow valve is designed for a specific flow rate at which it closes. The effect of piping, fittings, and valves downstream must be taken into account when evaluating flow. The valve should be installed as close as possible to the equipment being protected.

Valves should be selected with a closing flow rate of at least 10–15 percent greater than the anticipated normal flow. Valves having a rated closing flow nearer to the normal flow may chatter or slam closed when surges in the line occur during a normal operation, or due to the rapid opening of a control or quarter turn valve.

The closure speed of larger sized valves may be reduced by the addition of a dashpot which utilizes the working fluid. This is recommended only on large liquid lines, to avoid shock associated with sudden closure.

A downstream break in piping or hoses may not provide sufficient flow to close the valve. Therefore, as an alternative, supply lines may be fitted with a remote operated shut-off valve, operable from a number of points located at some distance from the line, such that access to at least one point is possible, irrespective of wind direction and prevailing conditions.

6.7 FLOAT OPERATED VALVES (Fig. 6.4)

Float operated valves are used for liquid level control in non-pressurized containers. Valves may be to BS 1212 parts 1, 2, 3 or to manufacturer's standards.

Valves to BS 1212 are small bore, sizes 3/8 in. to 2 in. NB, with threaded male end, specified with an orifice sized to accommodate various conditions of the pressure and flow. To enable the correct orifice size to be determined, the computed flow through each orifice at given heads is tabulated in part 1. Available orifice sizes are:

- part 1: $1/8$–$1\frac{1}{4}$ in;
- parts 2 and 3: $1/8$–$3/8$ in.

Valves to BS 1212 parts 2 and 3 have the outlet positioned above the body, rather than below as in part 1, enabling the attachment of a discharge assembly to prevent back siphonage of the fluid.

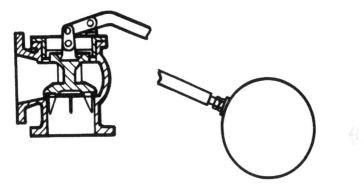

Fig. 6.4 Float operated valve
(*Courtesy of British Valve and Actuator Manufacturers' Association Ltd*)

Valves to manufacturer's standards range from the small threaded valve in the sizes covered by BS 1212, to flanged valves up to 18 in. NB, pressure balanced to equalize the hydraulic forces on the moving element and giving greater sensitivity to changes in water level throughout the inlet pressure range.

Valves to manufacturer's standard may be either in-line or angle type. Designs are available for high pressure applications, and have a streamline flow pattern which provides smooth handling of high velocities, minimizing vibration, erosion, and noise.

Surface turbulence, for instance in a break tank, can cause oscillatory action of the valve. This may be prevented by installing a separate float tank, or baffle plate.

This type of valve should always be backed up by some independent means of preventing tank overfilling, because of the valve's low reliability, even on low hazard duties.

6.8   FLUSH BOTTOM OUTLET VALVES (Fig. 6.5)

A flush bottom valve is a 'Y' pattern valve which controls the flow of liquid or slurry from the bottom of a vessel to a valve discharge angled at 45 degrees to the vertical and is generally to manufacturer's standards.

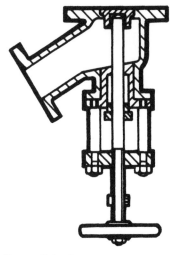

Fig. 6.5   Flush bottom outlet valve
(*Courtesy of British Valve and Actuator Manufacturers' Association Ltd*)

The valve may have a disk and seat, or be of seatless design with a piston-, plunger-, or mushroom-type disk. It may be selected for either flush bottom or penetrating operation.

A flush bottom valve allows removal of precipitate which may have bridged the vessel outlet and, when fitted to a dished end, allows complete draining of the vessel.

When heavy sediment may be deposited in the vessel, a piston- or plunger-type valve should be selected. On closing, the piston penetrates the deposit, allowing the vessel to drain when the valve is next opened.

Since the seat in a disk-type valve is part of the vessel outlet nozzle, the valve must be provided before the vessel is fabricated. Disk-type valves may not seal properly when used with liquids containing solids in suspension.

The valve requires a considerable vertical distance under the vessel bottom for installation and operation which may be manual or by remote control.

A variation of the piston design is used for sampling.

6.9   IRIS VALVES

These are valves in which the closure member moves towards the valve bore. They are mainly used for controlling powder media. There are several designs.

(a)   A flexible cylinder is rotated at one end, and closure is effected midway along the cylinder.
(b)   The closure member is in the form of flat petals hinged and rotated to close (like a camera diaphragm).
(c)   The petals are curved like a cone and hinged to close together; this design is used for quick shut-off on hydraulic systems where leakage is permissable.

6.10   ROTARY (OR METERING) VALVES

These valves consist of a spindle or hub to which several blades are attached, the whole being housed in a fabricated or cast body. Rotation of the spindle causes a measured quantity of product (powder or pellet) to be transferred from storage vessels to conveying lines. Valves are usually motorized and are not intended to seal tight. A specialized design of ball valve incorporating a cavity is also used for this duty.

6.11   EMERGENCY SHUTDOWN VALVES

Standard types of valve intended for ESD service must achieve the highest degree of reliability and integrity and are usually affected by legislation. The actuators of such valves are usually fail-safe hydraulic or pneumatic type and fundamentally affect the valve's ability to perform its shut-off function. In the case of offshore applications the Certifying Authority should be involved in all stages of the procurement process and specialist assistance should be sought.

The chief requirement is that the valve can be relied upon to close when called to do so under any likely conditions of operation. To demonstrate this, tests involving partial closure of the valve are sometimes required. Tight shut-off is generally a secondary consideration.

6.12 SUB-SEA VALVES

In sub-sea applications the utmost reliability is called for because of the very high cost of any intervention and the likelihood that export revenue will be affected if the valve has to be removed. Valves with top entry and single piece insertion of internals are often selected to facilitate maintenance which, in some cases, can be conducted *in situ* using divers or remote operated vehicles.

Sub-sea isolation valves are intended to prevent gross backflow of product from pipelines to offshore platforms in the event of a pipe rupture on the pipeline side of an ESDV. They are required to function for the lifetime of a facility with infrequent operation.

Materials of sub-sea valves must be selected with care if a long, maintenance-free life is to be achieved. Prior to operation, sub-sea systems are often left full of seawater for long periods and the effect of this on the valve must be considered. Overlay of seal pockets with inconel and the use of duplex steel where austenitic stainless material might otherwise be contemplated is strongly recommended. Elastomers, where used, should be chosen in full knowledge of the applicable service conditions (product composition, chemicals, depressurization rates, and so on) if expensive intervention is to be avoided.

Increased conservatism in design (e.g., operating torque/force requirements) will pay dividends, as will simplicity and robustness of construction. In the case of sub-sea choke valves use of the most erosion-resistant materials available, even when these are very expensive, is easily justified by the extended maintenance-free life which they can offer.

CHAPTER 7

Wellhead Gate Valves for the Petroleum Industry

This section provides a quick guide to assist in the selection of wellhead gate valves and considers:

- features of various types;
- valve selection;
- design requirements;
- materials;
- supporting calculations;
- testing.

7.1 VALVE TYPES

There are three principal types of gate valve which may be considered for use in wellhead applications.

7.1.1 Floating seat (Fig. 7.1)

The seats are spring loaded from the body such that both upstream and downstream seats are always in contact with the gate. In this design both the seat-to-gate and seat-to-body seals are dynamic seals.

7.1.2 Fixed seat/floating gate (Fig. 7.2)

The seats are fixed relative to the valve body, and the body-to-seat seal is provided by a static seal. Cavity pressure forces the gate to float and seal against the downstream seat. A floating connection is required between gate and stem. It is also possible to combine the best of both designs and use a fixed seat downstream and a floating seat upstream.

7.1.3 Split gate (Fig. 7.3)

In this design the gate is effectively spring loaded so as to contact both seats; the seats will generally be of the fixed variety. There are two basic types of split gate design.

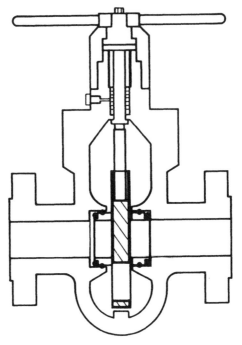

Fig. 7.1   Floating seat well head gate valve

(a) The first design, the spring loaded gate, maintains seat-to-gate contact.

(b) The second design is the spring retained expanding wedge design, in which the gate is constructed from two wedge halves. When one half contacts a stop in the body the other half is forced to ride up the inclined plane, thus forcing both gate halves out to form an interference fit with the seats and provide a seal.

7.2   VALVE SELECTION

The selection of an appropriate type of valve for a particular service must take account of the two following aspects.

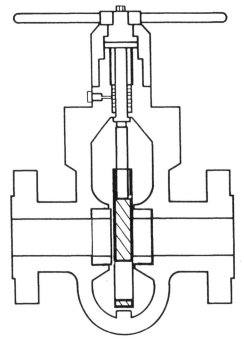

Fig. 7.2   Fixed seat well head gate valve

Selection considerations:

– actuation;
– sandy service.

7.2.1   Actuation

Valves which maintain seat-to-gate contact during operation must be designed to prevent cavity locking effects, i.e., an increase in cavity pressure which generates sufficient opposing force to prevent further movement. This can be achieved by the use of a balanced stem arrangement, i.e., an equal diameter stem above and below the gate.

The spring-retained expanding wedge design usually requires the generation of very high actuation forces to operate the valve. This generally

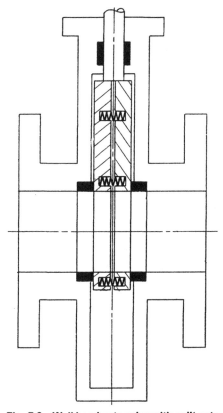

Fig. 7.3   Well head gate valve with split gate

precludes the use of a spring return actuator. For this reason this type of valve should not generally be used where fail safe closure is required (e.g., when control system pressure is lost).

Unbalanced cavity valve designs (i.e., single stem or two stems with different diameters) can produce an upwards pressure force to assist in valve closure, which reduces the size of actuator required. Such designs should prevent any potential change of pressure due to cavity volume changes by venting to the process bore.

7.2.2 Sandy service

Valves which maintain seat-to-gate contact are preferable for sandy service. Fully fixed seat designs can also be considered if the gate float tolerance is minimized.

Sandy service valves should feature sufficient debris seals and scrapers to prevent sand ingress into the gate, seat, and stem sealing faces. Seat plates are helpful in preventing communication between the process bore and the valve cavity during actuation of the gate of conduit-type valves.

7.3 DESIGN REQUIREMENTS

- Complexity and track record
- Stem seals
- Seat to gate sealing
- Seat to body sealing

7.3.1 General

The valve should be of minimum complexity and should either have a proven track record or have been previously rigorously tested.

7.3.2 Valve stem seals

Stem seals should, in general, consist of the following:

(a) a scraper to remove debris from the stem;
(b) a primary seal which should be metallic (The scraper and metal primary seal may be incorporated as one device.);
(c) a backup seal which should be constructed from a material highly resistant to chemical attack and explosive decompression damage (Typically the backup seal is constructed from glass-reinforced PTFE and incorporates a spring energiser for low pressure sealing.).

Many valve designs employ an arrangement where a profile on the stem can contact against a mating face in the bonnet in order to form a back seat. This arrangement is beneficial in two ways:

- it forms a further seal, possibly allowing changeout of the stem primary and backup seal whilst still under pressure;
- it prevents ejection of a disconnected stem out through the bonnet.

The use of elastomeric materials for stem seals is undesirable due to their relatively poor resistance to explosive decompression and attack from chemicals such as methanol and scale inhibitor. All seal material should be carefully selected to resist attack from any specified or likely process fluid.

The process fluid temperature and pressure will have a major effect on stem seal selection. PTFE glass-reinforced stem seals are generally suitable up to temperatures of 120°C combined with pressures up to 15 000 psi; a metal seal may have to be used for service above this limit. An extended bonnet design may be adopted in order to locate the stem seal in an area more remote from the high temperature fluid.

Fire certified valves will either need to have stem seals which are resistant to high temperature (e.g., metal) or seals which are shielded from high temperature effects.

The stem itself should be hard coated to avoid any scoring from debris or galling with the scraper or metal stem seal.

7.3.3　Seat-to-gate sealing

Floating or fixed seat designs are generally acceptable.

Gate-to-seat sealing in both high and low pressure should be effected by metal-to-metal contact only. The gate-to-seat contact pressure needs to be approximately three times that of the process fluid. Valves having soft insert seals in the seat sealing face are not recommended; such seals have only limited life and can often function as debris traps impairing valve sealing and causing wear.

The gate and seat surface finish and flatness should comply with the following requirements if low pressure sealing down to 5 bar is required.

(a)　a flatness better than three light bands should be obtained using a helium laser source;

(b)　the surface finish should be better than 8 micro-inches CLA.

In the case of a process fluid containing debris a sandy service-type design of valve should be utilized. The design must prevent damage from any particles being trapped between the downstream seat-to-gate sealing surface. This can be achieved by:

－　adopting spring loaded seats which always maintain contact with the gate;

- minimizing the gate float in the case of a floating gate design;
- adopting a split spring loaded gate design;
- using hard gate/seat materials.

Note that chamfers on the seat OD must be avoided since they can act as debris traps and cause scoring of the gate and seats.

7.3.4 Seat-to-body sealing

A metal-to-metal seal is often preferred in this location for both high and low pressure sealing.

In the case of a floating seat design the seal used should be capable of accepting repeated compression and expansion due to the seat movement. The sealing faces should also be protected by debris or scraper seals to prevent build up of sand etc., which could impair the float or sealing integrity.

7.4 MATERIALS

Material requirements will usually be covered in a technical specification (with reference to the working fluid).

In general, alloy steel will be suitable for the majority of applications. However, many services will require special consideration (e.g., where water is present or the product is sour) and overlay with nickel alloy (especially seat pockets) etc., or the use of duplex steels is becoming increasingly common.

The need for hard gate and seat materials cannot be over-emphasized. Typically gate and seats in the order of 50 HRC should be utilized, with a differential hardness of approximately 5 HRC between seats and gate. These hardness values will provide long-term wear resistance and avoid abrasion damage.

It is likely that coatings based on ceramic or diamond will come to be used more widely in future, especially for subsea applications.

Stems need to be constructed from a wear- and corrosion-resistant material to avoid damage from hard particles and to prevent any galling action with metal stem seals. Typically, a tungsten carbide coating will be required on top of a corrosion-resistant substrate material.

7.5   SUPPORTING CALCULATIONS

It is recommended that the manufacturer should be asked to provide the following:

(a)   calculations showing the sealing pressure developed by the seat-to-gate contact;

(b)   calculations of the maximum gate drag, stating what seat-to-gate coefficient of friction is used;

(c)   calculations justifying the pressure retaining boundary.

7.6   TESTING

Evidence of satisfactory testing of the valve design gives some confidence of satisfactory operation and should include:

- high pressure hydrostatic seat leak tests;
- low pressure hydrostatic seat leak tests;
- low pressure air tests at 5/6 bar;
- API 14D type test for sandy service conditions (if sandy service is required).

CHAPTER 8

Valve Materials

This section discusses materials suitable for some of the more commonly encountered services. The tables provide a general guide to the application of specific materials.

8.1 GENERAL

For most applications and non-corrosive duties, carbon steel is normally used for the pressure retaining boundary. For high temperature applications creep resisting alloy steel may be specified. Alloy steel, stainless steel, nickel alloys, plastics, rubber lined or other special materials may be required on corrosive services or where there are clean or other special requirements.

Materials for low temperature service (down to $-50°C$) may be carbon or alloy steel with impact test requirements. Aluminium or stainless steel may be specified for cryogenic service ($-50°C$ to $-196°C$). Flake graphite cast iron should be restricted to low pressure, non-hazardous applications and should be avoided where freezing may occur.

Materials used for valve trim should be suitable for exposure to the line fluid. Seating components may require to be manufactured from, or faced with, a hard material (e.g., nickel alloy, tungsten carbide, stellite, and so on), to withstand wear, erosion, and wire drawing. Performance of hard facings may depend on the suitability of the substrate material, particularly at low temperatures. Materials for valve stems should be chosen with a view to avoiding galling and corrosion when in contact with glands, trunnion bearings etc.

When choosing materials care must be taken to avoid galvanic action between dissimilar materials.

Tables 8.1–8.3 provide general application information for some of the more commonly encountered pressure boundary, trim, and non-metallic materials. These are typical recommendations only; where doubt or unusual conditions exist, specialist advice should be sought.

Table 8.1 Valve body and bonnet/cover materials

| Materials | Typical specifications | | Typical application |
	Forgings	Castings	
Carbon steel	BS 1503–221–490 ASTM A105	BS 1504–161–480 ASTM A216–Gr WCB	Non-corrosive process hydrocarbons, produced water, slurries, de-aerated sea water, water, air, steam
$1\frac{1}{4}$Cr$\frac{1}{2}$Mo steel	BS 1503–621–960 ASTM A182–Gr.F11	BS 1504–621 ASTM A217–Gr.WC6	HP steam and process. Good resistance to sulphur and hydrogen; good mechanical properties at elevated temperatures
5Cr$\frac{1}{2}$Mo steel	BS 1503–625–590 ASTM A1820–Gr.F5	BS 1504–625 ASTM A217–Gr.WC5	
$2\frac{1}{4}$ Cr1Mo steel	BS 1503–622–490 ASTM A182–F22	BS 1509–622 ASTM A217–WC9	High temperature power station steam service
Stainless steel type 304	BS 1503–304–S40 ASTM A182–Gr.F304	BS 1504–304–C15 ASTM A351–Gr.CF8	Corrosive service, low temperature service, services requiring cleanliness. Unsuitable for sea water service. Not to be used where chlorides exceed 30 ppm
Stainless steel type 316	BS 1503–316–S31/S33 ASTM A182–Gr.F316	BS 1504–316–C16 ASTM A351–Gr.CF8M	Highly corrosive service. Not recommended for sea water. Not to be used where chlorides exceed 30 ppm
Carbon steel (impact tested)	BS 1503–221–Gr.490 LT50 ASTM A350–Gr.LF2	BS 1504–161–Gr.480 LT50 ASTM A352–Gr.LCB	Low temperature service
Bronze		BS 1400 Gr.LG2 ASTM B62 Alloy B36	Black sewage, brine, firewater, air, steam. Leaded bronze has poor resistance to sea water

Material	Standard	Description
Aluminium bronze	BS 1400 Gr.AB2	Sea water, black sewage, brine, firewater. Good for high velocities. Unsuitable for sulphide polluted water
Titanium	ASTM B348 Gr.2	Sodium hypochlorite and ferritic chloride solutions
Grey cast iron (flake graphite)	BS 1452 Gr.220 ASTM A126 Class **B**	Land locations, water, aqueous solutions, non-volatile chemicals. Do not use for hydrocarbons or hazardous service. Should not be used where freezing may occur
Spheroidal graphite cast iron	BS 2789 ASTM A395	As grey cast iron, but may be used at higher pressures and temperatures
Monel 400	ASTM A494 or A744–M–35–1	Sea water, brackish water, brine. Good resistance to all acids except oxidizing types
Hastelloy alloy C	ASTM A494 or A744–CW–12M	Hypochlorites, acetic acid, chlorine, hydrogen
13% chrome steel	BS 1504–420–C29 ASTM A217–CA15	Natural gas + CO_2 (hardness limited to 22RC max)
Super Duplex steel	BS 1503–541–S21 ASTM A182–F60	Sea water service. Natural gas + extreme CO_2. Extreme sour service
UPBV PVDF PP ABS		Land locations or inside modules only. Non-fire hazardous services, water and process services

Table 8.2   Typical application of metallic trim materials

Material	Notes
13% chrome steel	General services, gases, oil, steam. Normally supplied with body materials LBC, WCB, WC1, WC6, WC9, C5, and C12. Note: stems may be subject to graphitic attack
13% chrome with nickel alloy facing	General services, steam, water, air, gas fuel oil, non-lubricating non-corrosive low viscosity oils. Normally supplied with WCB body material
13% chrome steel, hard faced	General services, steam, gas, oil, and oil vapour. Supplied with body materials LCB, WCB, WC1, WC6, WC9, and C5
Hard faced trim (e.g., stellite, tungsten carbide)	Steam, wire drawing applications, dirty service etc. Normally supplied with body materials WCB, WC1, WC6, WC9, C5, C12, CF8, CF8M, and CF8C
Stainless steel 18–10–2 with or without hard facing	Corrosive services. Normally supplied with body materials LC3, LC2, LC1, LCB, WCB, CF8, CF8M, and CF8C
Bronze	Cold/hot water, marine applications, and low temperature service. Normally supplied with WCB body material
Aluminium bronze	Sea water, brine, firewater. Unsuitable for sulphide polluted water
Super Duplex steel	Sea water, sour service
Hastelloy alloy C	Hypochlorites, chlorine, hydrogen sulphide, sea water, brine
Monel and inconel	Corrosive services
Electroless nickel plating	Used for ball valves
Cast iron	Not used where freezing is likely to occur
Titanium	Sodium hypochlorite

8.2   MATERIALS FOR FIRE HAZARD AREAS

Where fire is a possibility the following materials should not be used for valve components in flammable or toxic service:

- cast, malleable, wrought, or nodular iron;
- brittle, low melting point, or flammable materials such as aluminium, brass, or plastics.

Table 8.3   Typical application of non-metallic materials

Material	Application
Butyl rubber	Superseded for most applications by more advanced elastomers. Cold water service, good wear resistance, low oil permeability
Chlorinated polyether (Penton)	Process service (good resistance to acids and solvents). If softening can be tolerated can be used to 125°C
Chlorosulphonated polyethylene (Hypalon)	Good resistance to chemical attack (e.g., acids, alkalis, oxidizing agents, minerals, and vegetable oils), poor resistance to aromatic and chlorinated hydrocarbons
Ebonite	Good chemical resistance, most grades soften above 70°C. Becomes brittle at low temperatures
Ethylene propylenediene (EP DM)	Good mechanical properties, good resistance to phosphate ester-based hydraulic fluids and minerals. Good resistance to alcohols, ketones, and to weathering. Some grades can be considered for wellhead and hot water applications
Hydrogenated nitrile	Similar to nitrile but better in sour service and excellent explosive decompression resistance. Swollen by aromatics and adversely affected by amine corrosion inhibitors
Natural rubber	Suitable for cold water and some chemical and abrasive service. Has low resistance to solvents, oils, and sunlight
Nitrile rubber	Good general service material, good resistance to oil, solvents (but not aromatics), and chemicals. Subject to swelling when used with de-ionized water. Should not be used for sour service. Poor resistance to sunlight and weather
Nylon	Insoluble in hydrocarbons, good resistance to alkalis, will absorb water and swell. Good frictional properties, less prone to cold flow than PTFE
Polychloroprene (Neoprene)	Good resistance to most inorganic chemicals, refrigerants, water. Good resistance to sunlight and weather. Suffers from compression set when hot. Low resistance to aromatic hydrocarbons, oxidizing acids
Polypropylene	Good resistance to chemical attack. Similar to polythene but not subject to stress cracking
Perfluoroelastomer (kalrez, chemraz)	Almost universal application due to excellent chemical resistance. Mechanical properties not quite so good. Attacked by refrigerants. Extremely high cost owing to difficulties of manufacture
PEEK	Any application where tough, wear-resistant polymer with outstanding chemical resistance is required. Harder, tougher, and less prone to cold flow than PTFE but becomes brittle at low temperatures

(*Continued*)

Table 8.3   Typical application of non-metallic materials (*continued*)

Material	*Application*
Polyethylene	Good resistance to mineral acids, alkalis, and solvents. Suffers embrittlement when subject to polar solvents, esters, alcohols, and ketones
Polyurethane	Excellent resistance to oils, solvents, fats, grease, petrol, ozone, sunlight, and weather. Good properties at low temperatures. Some reduction in properties at high temperatures. Susceptible to hydrolysis, should not be used with hot water or acid. Will swell on contact with ketones, esters, aromatics
PTFE	Excellent for most process services with almost universal chemical resistance. Maximum temperature limits may be increased by adding fillers. Maximum allowable temperature depends on seal design. Low coefficient of friction, but subject to creep and cold flow under moderate loads
PVC	Can be supplied plasticized or in rigid form. Good chemical resistance. Can suffer from creep
Silicone rubber	Mainly static seals. Poor physical properties at room temperature, but well-maintained at low temperatures. Lack of resistance to chemical attack. Not resistant to acids or alkalis. Aromatic and chlorinated solvents and petrol cause swelling
Fluoroelastomer (Viton)	Water and process service. Good resistance to most chemicals including some acids, petrol, and solvents. Should not be used with esters and ketones. Poor flexibility at low temperatures. Viton A has poor methanol resistance. Viton GF is better, but has poor explosive decompression resistance. Filled grades can be highly resistant to ex. decomp.

Note: Most elastomers are subject to swelling when used with de-ionized water, particularly nitrile rubber. Viton and neoprene elastomers should not be considered for hot water applications because they are susceptible to blistering, but may be used for hot oil application. EPDM (peroxide cured) may be considered for hot water applications up to 130°C. It is recommended that the use of viton is limited to 180°C for pressurized systems and 200°C for non-pressurized systems.

8.3   ELASTOMERS AND PLASTICS

Some elastomers and plastics may be subject to swelling when used with particular fluids (e.g., nitrile rubber is not suitable in de-ionized water, nylon may swell with water). Elastomeric materials in contact with hydrocarbons should be of a type not prone to high gas permeation rates, which result in failure when subject to a sudden pressure drop (explosive decompression).

Valve pressure–temperature ratings are limited when non-metallic materials (PTFE, nylon, rubber) are used for seats, seals, linings, and gaskets.

For example, virgin PTFE is normally limited to a maximum operating temperature of 200°C but this may be increased marginally by the use of a suitable filler (e.g., glass). Guidance on temperature limits may be obtained from the valve or material manufacturer.

8.4 HIGH TEMPERATURE SERVICE

For high temperature (generally above 400°C) power station steam services $2\frac{1}{4}$%Cr 1%Mo steel is often chosen. The seats and disks will require hard facing with alloys of cobalt, chromium, and tungsten and a difference between the hardness of the disk and seat is recommended to avoid galling.

8.5 LOW TEMPERATURE SERVICE

At low temperatures (less than 0°C) most ferrous alloys suffer a reduction in notch impact strength and require Charpy impact testing to demonstrate adequate toughness.

Most non-ferrous materials such as copper, copper alloys, aluminium alloys, and nickel alloys are suitable for low temperature service. $2\frac{1}{2}$% nickel steels may be used down to -57°C, and austenitic stainless steels retain acceptable properties even under cryogenic conditions.

Cast iron should not be used for low temperature service.

8.6 ANHYDROUS AMMONIA SERVICE

Copper and copper-based materials should not be used in anhydrous ammonia service.

Steels should have a maximum yield strength of 350 N/mm^2 (e.g., low strength steel) and all welds in steel components should be stress relieved.

8.7   CHLORIDE SERVICE AND ENVIRONMENTS

Stainless steels may be susceptible to stress corrosion cracking in chloride-containing environments at temperatures above 60°C, depending on the concentration and stress level. For purposes of hydrostatic testing, the chloride content of the test water for valves containing stainless steel components (and which may be used at such temperatures) is best limited to 20 ppm – lower if evaporation and concentration is likely.

8.8   SOUR SERVICE

Materials for sour service should comply with the requirements of NACE standard MR–01–75 which limits the hardness of carbon steels, requires austenitic steels to be solution annealed, and provides special requirements for bolting, welding, and so on.

8.9   HYDROGEN SERVICE

Carbon steel may be used for hydrogen service up to a maximum of 230°C. The material should be selected in accordance with API standard 941 (Nelson curves); the choice depends on the mixture of free hydrogen and other fluids at particular hydrogen partial pressures and temperature.

8.10   WET CO_2 AND CHLORINE SERVICES

When choosing materials for wet CO_2 or wet chlorine services, specialist advice should be sought.

8.11   MATERIAL COMPOSITION OF WELDING END VALVES

The chemical composition (by ladle analysis) of valve bodies with butt weld or socket weld ends should be restricted in respect of carbon and carbon equivalent. The following rules should be applied:

(a) Carbon and carbon-manganese steels
Carbon $\leqslant 0.25\%$ (forgings and castings)
Carbon equivalent (CE) 2 $\leqslant 0.45\%$
where

$$(CE)2 = C + \frac{Mn}{6} + \frac{Cr + Mo + V}{5} + \frac{Cu + Ni}{15}$$

Note
If necessary, in order to achieve minimum specified strengths, it may be possible to increase the specified manganese contents, as given in the referenced specification, up to a maximum of 1.6 percent, provided the maximum permitted CE is not exceeded.

(b) Chromium–molybdenum steels
Steels with 2% Cr and less:

carbon $\leqslant 0.20\%$ (forgings and castings)

Steels with more than 2% Cr:

carbon $\leqslant 0.15\%$

Residual elements:

copper $\leqslant 0.30\%$
nickel $\leqslant 0.40\%$
tungsten $\leqslant 0.10\%$
tin $\leqslant 0.03\%$

Austenitic stainless steel valves which are to be welded should be manufactured from a low carbon grade (e.g., 304L, 316L) or stabilized grade (e.g., 321) in order to avoid corrosion resulting from the sensitization of the material.

8.12 PLATED COMPONENTS

When specifying valves (especially ball valves) for corrosive service (e.g., produced water) it is necessary to take into account the fact that plating is usually porous so that, if problems are to be avoided, the base material must be corrosion-resistant (e.g., stainless steel). The following guidelines are suggested for electroless nickel plate:

(a) minimum plating thickness should be 0.08 mm;
(b) plating should be free of porosity;
(c) phosphorous content should be at least 10 percent;
(d) plating should have a maximum of 0.05 percent of elements other than nickel and phosphorous;
(e) the heat treatment should be conducted in accordance with a written procedure;
(f) the base material should be compatible with the plating to ensure adequate bonding.

Plating is rarely effective where the service is abrasive and overlaid coatings (e.g., tungsten carbide) are generally preferred for such applications.

8.13   GLAND PACKINGS

Compression packings consist of deformable materials such as asbestos fibres, exfoliated graphite, or polymer filament fibres/yarn. The material is typically in the form of a box section, in a continuous coil, or separate preformed ring elements; the latter are preferred. Compression by the gland follower urges the packing against the valve stem and stuffing box wall to provide a seal. Such packings traditionally suffer from relaxation of the load in service which causes leakage. To combat this, and where sealing integrity is of prime importance, techniques such as live (spring) loading and chamfered packing rings may be employed.

Graphite packing containing even small amounts of impurities can cause corrosion of martensitic stainless steel valve stems in the presence of water (e.g., after hydrostatic testing). To avoid this, such material should be supplied with a corrosion inhibitor having a passivating or sacrificial action.

A common cause of leakage is overlong stuffing boxes with too many packing rings such that the load applied by the gland follower never reaches the lower rings. Reducing the length of such arrangements (e.g., by replacement of superfluous packing by a rigid spacer) can often be beneficial.

Chevron packings generally consist of 'V' shaped rings of PTFE and reinforced polymers, which are preloaded by the gland and pressure energized by the process media. They are most frequently used in packed glands where friction must be minimized (e.g., control valves).

'O' ring seals consist of round section elastomer rings fully retained in properly designed housings and sealed by pressure energy from the process media. Materials must be carefully selected to suit the fluid being handled. Specially filled grades are available for use in gas service where explosive decompression is a possibility.

Reinforced lip seals consisting of an outer sheath of polymer (e.g., PTFE) with internal nickel or stainless steel reinforcement are also pressure energized. They can be very effective but will leak if scratched across the sealing face.

Thrust packings are packing rings or washers of polymeric reinforced materials mounted on shoulders in the bonnet and on the valve stem; initial sealing may be required by a compression packing. They are only used with quarter turn valves.

Diaphragm seals of diaphragm valves are used to isolate the valve stem from the process media. Various designs are used including bellows diaphragms of elastomeric or polymeric materials. It is important to realize that, unless a secondary seal (e.g., 'O' ring) is provided on the stem, diaphragm rupture will result in leakage to atmosphere.

8.14 BOLTS, NUTS, AND SCREWS

Material for bolts, studs, screws, and nuts should be selected to suit the flange materials and conditions of service. For sour service 'M' grade bolting will be required and for low temperature applications, impact tested alloy seal material (L grade) should be specified. For cryogenic service, austenitic stainless steel is frequently used but it should be remembered that this material is much weaker than alloy steel, so bolt load may be restricted where a direct substitution is made without redesign of the joint.

Bolts for use on offshore applications should be zinc or cadmium plated. PTFE coating can also be effective but, if the coating is ruptured, corrosion may be accelerated.

8.15 MATERIAL TEMPERATURE LIMITATIONS

The maximum and minimum operating temperatures given in Tables 8.4–8.6 are a general guide only for non-corrosive conditions. The corrosive

Table 8.4   Temperature limits of body and bonnet cover materials

Materials	Temperature °C	
	Min	Max
Carbon steel	−29	425
Chromium molybdenum (1¼Cr, ½Mo)	−29	593[1,2]
Chromium molybdenum (5Cr, ½Mo)	−29	593[1]
Stainless steel type 304	−196	538
Stainless steel type 316	−196	538
Carbon steel (impact tested) LT50/LF2	−50	343
Bronze	−30	260
Aluminium bronze	−30	260
Titanium	−30	315
Grey cast iron	−	204
Spheroidal graphite cast iron	−	343
Monel 400	−196	425
Hastelloy C	−196	425
13 chrome steel	−50	600
Duplex stainless steel	−50	315

[1] Class 150 flanges/valves 540°C max
[2] Scaling may occur above 565°C
All valve temperature limits may have to be reduced to suit trim and
other materials.

Table 8.5   Temperature limits of metallic trim materials

Materials	Temperature °C	
	Min	Max
13% chromium steel	−50	600
13% chromium with nickel alloy or other hard facing	−50	450–600 (depending on facing material)
Stellite or tungsten carbide hard facing	−196	650
Stainless steel 18–10–2 with or without hard facing	−196	450
Bronze	−196	288
Aluminium bronze	−196	260
Monel	−196	425
Inconel	Dependent on grade	
Duplex stainless steel	−50	315
Hastelloy alloy C	−196	425

Table 8.6 Temperature limits of non-metallic materials

Materials	Temperature °C*	
	Min	Max
Butyl rubber	−50	120
Chlorinated polyether (Penton)	90	
Chlorosulphonated polyethylene (Hypalon)	−15	200
Ebonite	0	57–149
		(depends on grade)
Perfluoroelastomer (kalrez etc.)	0/35	230/260
		(depends on grade)
PEEK	−10	250
Ethylene propylene (EPDM)	−30	150
Natural rubber	−50	70
Polychloroprene (neoprene)	−20	100
Nitrile rubber	−20	120
Polypropylene	0	100
Polyurethane	−30	90
Polyethylene	0	60
PVC	−40	60
Silicone rubber	−80	170
PTFE	−190	260
Fluoroelastomer (Viton)	(−5 to −35)	200
		(depends on grade)
Hydrogenated nitrile	−20	150

* For continuous exposure. Most materials will withstand brief excursions to 5° lower and 10/20° higher. For marginal applications specialist advice should be sought.

nature or condition of the fluid may restrict the allowable operating temperature range and service life of the material. Where any doubt of the material suitability exists, a materials specialist should be consulted.

8.16   CHEMICAL, ETC., RESISTANCE CHARTS (TABLES 8.7 AND 8.8)

Tables 8.7 and 8.8 are a guide for choosing valve materials. The choice is influenced by a number of factors such as fluid concentration, temperature, line velocity, presence of impurities, abrasion, possibility of stress corrosion cracking, and so on. Where any doubt exists, a materials specialist must be consulted.

Table 8.7 Elastomers

Property	Elastomer						
	Natural rubber	Neoprene	Nitrile	Butyl	Hypalon	EPDM	Viton
Abrasion resistance	A	A	B	BC	B	B	B
Gaseous impermeability	C	C	C	A	B	C	A
Tear resistance	AB	B	B	B	B	D	B
Cold resistance	B	B	B	B	BC	AB	BC
Resilience	A	A	C	D	C	C	D
Flame resistance	D	B	D	D	B	D	B
Ozone resistance	C	A	C	A	A	A	A

Key: A = Excellent; B = Good; C = Fair; D = Poor

Table 8.8 Chemical resistance of metallic and non-metallic material

Fluid	Cast iron	Carbon steel	304 SS	316 SS	Alum. bronze	Bronze	Monel	Hastelloy C	Titanium	13Cr SS	Cobalt based HF	22% Cr Duplex	17–4 PH SS	UPVC	PVDF	PP	ABS	PTFE	Nylon	Natural rubber	Neoprene	NBR (Nitrile)	Butyl rubber	Hypalon	EPDM	FKM (viton)
Acetic acid 10%	D	D	C	A	B	D	B	A	A	D	A	–	B	A	A	A	A	A	D	B	D	B	A	B	A	B
Acetone (10%)	B	B	B	B	B	A	A	A	A	A	A	–	A	D	B	A	D	A	A	B	A	A	A	B	A	B
ATP	A	A	A	A	A	A	A	A	A	A	A	A	A	D	D	D	D	A	A	B	A	A	A	A	D	A
Alums. (10%)	C	D	B	B	C	C	B	B	B	–	A	–	C	A	A	A	A	A	A	A	A	A	A	A	A	A
Amines	C	A	A	A	D	D	A	A	–	A	A	A	A	D	D	D	D	A	A	–	–	C	–	C	B	D
Ammonia aqueous	A	A	A	A	D	D	C	B	A	A	A	A	A	–	A	A	A	A	A	B	–	B	B	B	–	B
Aviation fuel	D	A	A	A	D	D	A	A	A	A	A	A	A	D	D	D	D	A	A	D	D	A	D	D	D	A
Barytes	A	A	A	A	A	A	A	A	A	A	A	A	A	D	D	D	D	A	–	A	A	A	A	A	A	A
Biocide	D	D	D	C	C	D	C	B	A	D	–	–	D	A	B	B	A	A	D	D	A	D	A	A	A	D
Brines	C	C	B	B	B	B	B	B	A	B	A	A	A	A	A	A	–	A	A	A	A	A	A	A	A	B
Butane gas	D	A	A	A	D	D	A	A	A	A	A	A	A	D	D	D	D	A	A	D	A	A	D	B	D	A
Calcium chloride	D	D	D	C	C	C	A	A	A	D	–	–	D	A	A	A	A	A	A	A	A	A	A	A	A	A
Calcium hypochlorite (2%)	D	D	D	C	D	D	D	B	A	D	–	–	D	A	A	A	A	A	A	D	D	A	A	A	A	A
Carbonic acid	D	D	B	B	B	D	D	A	A	A	A	B	A	A	A	A	–	A	A	A	A	A	A	A	A	A
Carbon dioxide (dry)	C	A	A	A	A	A	A	A	A	A	A	A	A	A	A	A	A	A	A	–	A	A	–	A	–	A

(Continued)

Table 8.8 Chemical resistance of metallic and non-metallic material (*continued*)

Fluid	Cast iron	Carbon steel	304 SS	316 SS	Alum. bronze	Bronze	Monel	Hastelloy C	Titanium	13Cr SS	Cobalt based HF	22% Cr Duplex	17-4 PH SS	UPVC	PVDF	PP	ABS	PTFE	Nylon	Natural rubber	Neoprene	NBR (Nitrile)	Butyl rubber	Hypalon	EPDM	FKM (viton)
Carbon dioxide (wet)	D	D	A	A	B	B	A	A	A	A	A	A	A	A	A	A	A	A	A	A	A	A	A	A	A	A
Carbon disulphide	D	B	B	B	D	D	C	B	A	B	A	–	B	D	D	D	D	A	C	D	D	D	D	D	D	A
Chlorine gas (moist)	D	D	D	D	D	D	D	B	A	D	B	–	D	B	D	D	D	A	D	D	D	B	D	D	D	A
Condensate (steam)	A	A	A	A	C	B	A	A	A	A	A	A	A	D	D	D	D	A	–	D	D	D	B	B	B	D
Copper sulphate (0–100%)	D	D	B	B	B	D	D	D	B	B	A	–	–	A	A	A	A	A	A	A	A	A	A	A	A	A
Ethane	D	A	A	A	D	D	A	A	A	A	A	A	A	D	D	D	D	A	A	D	A	A	D	B	D	A
Ethers	B	B	A	A	A	A	A	B	B	A	A	A	A	–	–	–	–	A	C	C	D	C	C	D	D	D
Ethylene glycol	B	B	A	A	A	A	A	A	A	A	A	A	A	A	A	A	A	A	A	A	C	A	A	A	A	A
Fatty acids	D	C	B	B	B	B	A	A	B	–	A	–	A	A	A	A	–	A	A	–	B	B	D	D	D	A
Ferric chloride (5%)	D	D	D	D	D	D	D	B	A	D	C	–	D	–	–	–	D	A	A	A	A	A	A	A	A	A
Ferrous sulphate	D	D	B	B	B	D	D	B	B	A	–	C	–	B	–	–	–	A	A	A	A	A	A	A	A	A
Foam (fire)	A	A	A	A	A	A	A	A	A	A	A	A	A	A	A	A	A	A	A	A	A	A	A	A	A	A
Formaldehyde (hot) (40%)	D	D	B	B	B	B	B	B	B	A	–	B	–	B	D	A	D	A	A	A	–	B	B	A	B	A
Formaldehyde (cold) (40%)	D	D	B	B	B	B	B	B	B	A	B	B	–	B	A	A	A	A	A	A	A	A	A	A	A	A
Formic acid (0–50%)	D	D	D	C	C	C	C	A	C	D	B	–	B	A	A	A	A	A	D	D	A	D	A	A	A	A
Freeon (dry)	B	B	A	A	A	A	A	A	A	A	A	A	A	–	A	B	–	A	A	D	A	A	B	B	C	C
Gas, condensate	D	A	A	A	D	D	A	A	A	A	A	A	A	D	D	D	D	A	A	D	B	A	D	B	D	A
Gas, fuel	D	A	A	A	D	D	A	A	A	A	A	A	A	D	D	D	D	A	A	D	B	B	D	B	D	–
Gas, inert	A	A	A	A	A	A	A	A	A	A	A	A	A	A	A	A	A	A	A	A	A	A	A	A	A	A
Gas, liquified petroleum	D	A	B	B	B	D	D	A	A	A	A	A	A	A	D	D	D	D	A	A	D	B	A	D	D	A
Gas, natural	D	A	A	A	D	D	A	A	A	A	A	A	A	D	D	D	D	A	A	D	A	A	D	A	D	A
Gas, produced	D	A	A	A	D	D	A	A	A	A	A	A	A	D	D	D	D	A	A	D	A	B	D	B	D	A
Gas, sour	D	A	A	A	D	D	A	A	A	A	A	A	B	D	D	D	D	A	A	D	A	B	D	B	D	A
Glycols	B	A	A	A	A	A	B	A	A	–	A	A	–	A	A	A	C	A	B	A	A	A	A	A	A	A
Halon	A	A	A	A	A	A	A	A	A	A	A	A	A	–	–	–	–	–	A	A	A	A	–	–	–	–
Helium	B	–	A	A	D	A	–	A	A	A	A	A	–	–	–	–	–	A	–	A	A	A	A	A	–	A
Hydrogen	A	A	A	A	A	A	A	A	A	A	A	A	A	A	A	A	A	A	A	B	A	A	A	A	A	A
Hydrochloric acid (to 30%)	D	D	D	D	D	D	D	B	D	D	B	D	D	A	A	A	A	A	D	A	D	D	A	B	A	C
Hydrofluoric acid (conc.)	D	D	D	D	B	D	B	B	B	D	D	C	D	D	A	A	A	A	A	D	D	D	D	A	D	C

(*continued*)

Table 8.8 Chemical resistance of metallic and non-metallic material (*continued*)

Fluid	Cast iron	Carbon steel	304 SS	316 SS	Alum. bronze	Bronze	Monel	Hastelloy C	Titanium	13Cr SS	Cobalt based HF	22% Cr Duplex	17–4 PH SS	UPVC	PVDF	PP	ABS	PTFE	Nylon	Natural rubber	Neoprene	NBR (Nitrile)	Butyl rubber	Hypalon	EPDM	FKM (viton)
Hydrogen peroxide (dilute)	D	D	B	B	C	D	B	A	A	C	–	–	B	A	A	A	A	A	D	B	D	B	A	A	A	A
Hydrogen peroxide (conc.)	D	D	B	B	D	D	B	B	B	–	–	–	B	–	A	D	D	A	–	B	D	D	B	B	B	B
Hydrogen sulphide	C	C	C	B	C	C	C	B	B	D	B	–	D	A	A	A	A	A	–	C	B	D	A	A	A	A
Kerosene	D	A	A	A	D	D	A	A	A	A	A	A	A	D	D	D	D	A	A	D	B	A	D	B	D	A
Methane	D	A	A	A	D	D	A	A	A	A	A	A	A	D	D	D	D	A	A	D	D	A	D	B	D	A
Methyl alcohol (0–100%)	B	B	B	A	B	B	A	A	D	A	A	–	A	A	A	A	A	A	A	A	A	A	A	A	A	C
Mud, drilling	B	A	A	A	B	B	A	A	A	A	A	A	A	A	A	A	A	A	A	A	A	A	A	A	A	A
Naphthalene	D	A	A	A	D	D	B	B	B	A	B	B	A	B	D	D	D	D	A	A	D	D	A	D	D	A
Nitrogen	A	A	A	A	A	A	A	A	A	A	A	A	A	A	A	A	A	A	A	A	A	A	A	A	A	A
Oil, crude (sweet)	D	B	A	A	D	D	A	A	A	A	A	A	A	D	D	D	D	A	A	D	B	A	D	B	D	A
Oil, crude (sour)	D	C	A	A	D	D	A	A	A	A	A	A	B	D	D	D	D	A	–	D	B	A	D	B	D	A
Oil, diesel fuel	D	A	A	A	D	D	A	A	A	A	A	A	A	D	D	D	D	A	A	D	B	A	D	D	D	A
Oil, hydraulic	D	A	A	A	B	B	A	A	A	A	A	A	A	D	D	D	D	A	A	D	A	A	C	C	C	A
Oil, lubricating	D	A	A	A	B	B	A	A	A	A	A	A	A	D	D	D	D	A	A	D	A	D	A	D	A	A
Oil, petroleum (refined)	D	A	A	A	D	D	A	A	A	A	A	A	A	D	D	D	D	A	A	D	B	A	D	B	D	A
Oil, petroleum (sour)	D	A	A	A	D	D	A	A	A	A	A	A	B	D	D	D	D	A	–	D	B	A	D	B	D	A
Oleic acid	C	C	B	B	C	C	A	A	A	A	A	–	B	A	A	A	A	A	B	D	D	B	D	D	A	A
Oxygen	D	A	A	A	D	D	A	A	A	A	A	A	A	D	D	D	D	A	–	D	A	D	A	A	A	A
Potassium carb. (aqueous)	B	B	B	B	B	B	B	B	A	–	B	–	B	A	A	A	A	A	A	–	A	A	B	A	A	C
Potassium chlor. (0–10%)	D	D	C	B	C	C	A	A	A	D	–	–	C	A	A	A	A	A	A	–	A	A	A	A	A	A
Propane	D	A	A	A	D	D	A	A	A	A	A	A	A	D	D	D	D	A	A	D	A	A	D	B	D	A
Sewage	C	C	B	B	B	C	C	B	–	B	A	B	B	–	–	–	C	A	–	A	A	A	A	–	A	–
Sodium bisulphite (<100%)	D	D	B	B	B	C	B	B	B	–	B	–	C	A	A	A	–	A	A	A	A	B	A	A	A	B
Sodium chloride	C	C	B	B	A	A	A	A	A	B	A	–	B	A	A	A	A	A	A	A	A	A	A	A	A	A
Sodium chromate (0–10%)	A	A	A	A	C	C	A	A	A	A	A	–	A	A	A	A	A	A	A	A	A	A	A	A	A	A
Sodium hydroxide (<40%)	B	A	B	A	B	D	A	A	A	B	A	–	A	A	D	A	A	A	C	C	A	A	A	A	A	B

(*continued*)

Table 8.8   Chemical resistance of metallic and non-metallic material (*continued*)

Fluid	Cast iron	Carbon steel	304 SS	316 SS	Alum. bronze	Bronze	Monel	Hastelloy C	Titanium	13Cr SS	Cobalt based HF	22% Cr Duplex	17-4 PH SS	UPVC	PVDF	PP	ABS	PTFE	Nylon	Natural rubber	Neoprene	NBR (Nitrile)	Butyl rubber	Hypalon	EPDM	FKM (viton)
Sodium hypochlorite (7%)	D	D	D	C	C	D	C	A	A	D	-	-	C	A	A	B	A	A	C	C	D	D	B	A	B	A
Sodium sulphite (25%)	B	B	B	B	D	D	B	B	A	-	B	-	B	A	A	A	A	A	A	-	A	A	A	A	A	A
Steam	B	B	A	A	C	B	A	A	A	B	A	A	A	D	D	D	D	A	-	D	D	D	B	D	B	D
Sulphur dioxide (wet)	D	D	C	B	B	B	C	B	B	-	B	-	B	A	A	A	D	A	-	D	D	D	B	D	A	A
Toluene	A	A	A	A	A	A	A	A	A	A	A	A	A	D	A	A	D	A	A	D	D	D	D	D	D	A
Tributyl phosphate	B	A	A	A	B	B	A	A	-	A	A	A	-	D	A	A	-	A	-	B	D	B	B	D	A	D
Water, chlorinated	D	D	D	C	D	D	D	B	A	-	-	-	-	B	A	B	A	B	-	A	B	D	B	B	B	B
Water, demin.	D	D	A	A	A	A	A	A	A	B	A	A	-	A	A	A	A	A	A	A	A	A	A	A	A	A
Water, potable	C	B	A	A	A	A	A	A	A	A	A	A	A	A	A	A	A	A	-	A	A	A	A	A	A	A
Water, produced	D	B	B	B	B	D	B	A	A	A	B	A	A	A	A	A	A	A	A	B	B	D	B	B	B	A
Water, sea (chlorinated)	D	D	D	C	D	D	D	B	A	D	A	B	D	B	A	B	A	B	-	A	B	D	B	B	B	B
Water, sea (deaerated)	B	B	B	B	A	A	A	A	A	B	A	B	B	A	A	A	A	A	A	A	B	A	B	B	A	A
Water, sea (raw)	B	B	C	B	A	A	A	A	A	D	A	B	D	A	A	A	A	A	A	B	B	A	B	B	A	A
Water, sea (polluted)	B	B	C	B	D	B	A	A	A	D	A	B	D	A	A	A	A	A	A	A	-	A	A	B	-	-
Zinc bromide	D	D	D	C	D	D	C	A	A	-	-	-	-	A	A	A	-	A	-	-	A	A	D	A	A	A

Key: A: Excellent resistance   B: Fair to good resistance   C: Poor resistance
         D: Not recommended
Abbreviations:
UPVC = Unplasticized polyvinyl chloride   ABS = Acrylonitrile–butadiene–styrene
EPDM = Ethylene propylene rubber   PVDF = Polyvinylidenefluoride
FKM = Fluorinated rubber (e.g., viton)   PP = Polypropylene
NBR = Nitrile rubber   PTFE = Polytetrafluorethylene
SS = Stainless steel
Notes:
1 The information given in this table is included as a general guide to the chemical resistance of the particular materials listed. Fluid temperature, pressure, concentration and
   material hardness may alter the suitability.
2 Fluids listed have resistance based on a temperature of 20° and are in a saturated/pure
   condition unless otherwise stated.
3 Irrespective of chemical resistance properties, cast iron, copper alloy, or plastic valves
   should not be used on hydrocarbon, toxic, or other hazardous service. Accordingly,
   resistance symbol 'D' has been used for these materials when listed against hazardous
   services.

CHAPTER 9

Sizing and Resistance to Flow

Depending on the type involved, valves can be a major source of pressure loss in piping systems. In this chapter simple calculations are given for:

incompressible (liquid) flow;
compressible (gas or vapour) flow;
surge.

9.1 INCOMPRESSIBLE (LIQUID) FLOW

There are four commonly used methods of stating a valve's resistance to flow.

(a) *Loss coefficient*, K

This is the most convenient form for use in piping system design since it can be added to the loss coefficients of other piping components and the pipe itself to produce an overall loss coefficient for the system. Pressure or head loss can be calculated as follows:

Head loss $\quad \Delta H = \dfrac{KV^2}{2g}$

Pressure loss $\quad \Delta P = \dfrac{KV^2 \rho}{2}$

where $\quad V =$ velocity of flow in pipe

$\qquad \rho =$ density of fluid

$\qquad g =$ acceleration due to gravity

(b) L/D *ratio*

This is the equivalent length of pipe, L, having internal diameter D which has the same resistance as the valve.

$$\text{Loss coefficient } K = \frac{fL}{D}$$

where f = pipe friction coefficient
L = pipe length
D = pipe diameter

A valve resistance quoted in terms of L/D ratio implies that an assumption has been made regarding pipe roughness and Reynolds Number and hence the value of 'f'.

(c) *Valve flow coefficient,* C_v
This is the flow of water (at 60°F), measured in US gall/min which can be passed by the valve when a pressure drop of 1 psi acts across it. Pressure drop under flowing conditions is then given by

$$\Delta P = \frac{Q^2 G}{C_v{}^2} \text{ lbf/in}^2$$

where Q = flow rate (US gall/min)
G = specific gravity of liquid

Note: If C_v is quoted in Imperial gallons the flow rate will need to be adjusted accordingly and vice versa. If flow is not fully turbulent (e.g., highly viscous liquid) a viscosity correction factor must be applied to C_v.

(d) *Metric valve flow coefficient,* K_v
This is the metric equivalent of C_v and represents the flow in cubic metres/hr which can be passed by the valve when a pressure drop of 1 bar (actually 0.981 bar) acts across it. Pressure drop under flowing conditions is then given by

$$\Delta P = \frac{Q^2 G}{K_v{}^2}$$

where Q = Flow rate (cubic metres/hr)
G = Specific gravity of liquid

The relationship between C_v and K_v is

$K_v = 0.87C_v$ (for C_v in US gallons)

$K_v = 1.05C_v$ (for C_v in Imperial gallons)

The relationship between C_v and K is

$K = \dfrac{894d^4}{C_v^2}$ (for C_v in US gallons)

$K = \dfrac{620d^4}{C_v^2}$ (for C_v in Imperial gallons)

(where d is in inches).

Table 9.1 provides guidance on the relative flow resistance of different valve types assuming fully turbulent flow. In specific cases, accurate information should always be obtained from the manufacturer.

The increase in velocity through a valve may result in the vapour pressure of the fluid being reached, causing cavitation when the pressure rises again with resulting erosion damage. Such situations should be avoided.

Table 9.1   Loss coefficient K

Valve type	Approximate K value	
	Min.	*Max.*
Globe and Lift check, 2 in. and above, full bore	4	10
Globe and Lift check ,$1\frac{1}{2}$ in. and below, full bore	5	13
Globe 45 degrees oblique type, full bore	1	3
Globe angle pattern 2 in. and above, full bore	2	5
Globe angle pattern $1\frac{1}{2}$ in. and below, full bore	1.5	3
Gate valve, full bore	0.1	0.3
Ball valve, full bore	0.1	–
Plug valve (rectangular port), full open	0.3	0.5
Plug valve (rectangular port), 80 percent open	0.7	1.2
Plug valve (rectangular port), 60 percent open	0.7	2.0
Plug valve (circular port), full bore	0.2	0.3
Butterfly valve	0.2	1.5
Diaphragm (weir type)	2.0	3.5
Diaphragm (straight through type)	0.6	0.9
Swing and tilting disk check	1.0	–

9.2 COMPRESSIBLE (GAS OR VAPOUR) FLOW

The situation here is much more complex than in the case of incompressible flow and there is less agreement on procedures. Valve manufacturers use a variety of different formulae for sizing, many of which involve the use of empirically derived factors. The user is primarily interested in the total pressure drop across the valve but the condition within the valve itself (e.g., possible high pressure drop followed by recovery, choked flow, and so on) must be taken into account in deriving this. In the case of full bore, low pressure drop valves (ball, gate) such effects are unlikely to be significant, but in other cases it is recommended that accurate information is obtained from the manufacturer and that authoritative texts be consulted (e.g., *ISA Handbook of Control Valves*, J.W. Hutchinson; *Compressible Internal Flow*, D.S. Miller (BHRA)).

The following formula is provided for guidance.

Loss Coefficient K (see Table 9.1)

$$K = 2\left[\frac{A_a}{A_v}\right]^2$$

where $A_v = 24 \times 10^{-6} \times C_v$ (or $28 \times 10^{-6} \times K_v$)

$A_a = 24.4 \times 10^{-3} \times A_v \times C$ (for inlet velocities of Mach 0.2 or less)

$$A_a = 24.4 \times 10^{-3} \times A_v \times C\left[1 + \frac{(\gamma + 1)}{2}M_1^2\right]^{-\frac{\gamma+1}{2(\gamma-1)}}$$

(for inlet velocities > Mach 0.2) where γ = ratio of specific heats and M_1 = Mach number at inlet.

C = Gas flow factor from manufacturer (based on air tests with critical flow). This is sometimes combined with C_v and quoted as a gas sizing coefficient Cg.

9.3 SURGE

Valves can be a source of surge phenomena such as 'water hammer' in a piping system. The usual cause is fast closure (e.g., a swing check valve

**Table 9.2   Approximate K factors
for axial flow, spring-operated,
non-slam check valves**

Size NPS	K value
2	1.9
3	1.7
4	1.6
6	1.4
8	1.3
10	1.2
12	1.2
14	1.1
16	1.1
18	1.1
20	1.0
24	1.0
28	0.97
30	0.95
32	0.94
36	0.92
40	0.89
42	0.89
48	0.86
52	0.85
60	0.82

slamming shut on flow reversal) but fast opening of a valve isolating a high energy line from a depressurized part of the system can also be problematic.

The effect is a rapid conversion of kinetic energy into strain energy (pressure rise and deformation of the piping system) and specialist assistance should be sought if such situations are likely to arise.

As an indication of where this might be necessary, valve closures occurring within the time taken for a pressure wave to travel from the valve to the other end of the pipeline and back (one wave cycle) must be regarded as occurring instantaneously.

$$\text{Time for one wave cycle} = \frac{2L}{C}$$

where L is the length of pipe and C is the wave velocity.

As will be seen from Table 9.3 below, the latter quantity is a function of pipe diameter and wall thickness as well as the fluid contained.

Table 9.3 Approximate wave velocities for instantaneous closure

d/t	C *wave velocity* (ft/s)	(m/s)
20	4300	1310
40	4000	1219
60	3800	1158
80	3600	1097
100	3400	1036
150	3100	945
200	2800	853
250	2600	792
300	2400	732

d = pipe inside diameter (inches or mm)
t = pipe wall thickness (inches or mm)

APPENDIX A

Valve Selection Tables

To assist in the selection of probable valve types for the majority of general services, a number of tables have been included.

These tables are for guidance only and the user should always ascertain that the service conditions are within the valve manufacturer's recommendations. They should be regarded as a supplement to individual experience of particular services and products.

Tables A1–A3 and A6–A7 may be used to obtain a recommendation for a valve or valves based on the appropriate conditions or required size and features selected. Tables A4 and A8 give guidance on materials availability and Table A5 indicates the degree of standard leak tightness to be expected from new valves.

In some cases the tables suggest that a variety of valve types are suitable, and the user may consider past experience for the service together with other factors e.g., if slow or quick opening/closing action is required (gate or ball valves) or relative cost. Further assistance can be obtained from more detailed information given elsewhere.

In general, the more benign the service, the more potentially suitable valve types there will be. A couple of examples are provided to illustrate the use of these tables.

Table A1	Selection of isolation (block) valves – service conditions
Table A2	Selection of isolation (block) valves – service conditions
Table A3	Selection of isolation (block) valves – features
Table A4	Selection of isolation (block) valves – available materials
Table A5	Selection of isolation (block) valves – achievable leakage rates
Table A6	Selection of check and diverter valves – service conditions
Table A7	Selection of check and diverter valves – service conditions and sizes
Table A8	Selection of check and diverter valves – available materials

An Introductory Guide to Valve Selection

Table A1 Selection of isolation (block) valves – service conditions

Valve types		Wedge gate				Parallel gate					Ball metal seat			Ball soft seat			Plug taper				
	Conditions	Solid wedge	Flexi-wedge	Split-wedge	Rubber lined	Conduit slab gate	Parallel double disk gate	Conduit split gate	Parallel slide (steam/feed)	Knife gate	Floating ball	Trunnion mounted ball	Eccentric ball	Floating ball	Trunnion mounted ball	Eccentric ball	Non-lubricated (sleeved)	Non-lubricated (lined)	Lubricated	Lubricated (balanced plug)	Lifting plug
Resistance to flow	Very low resistance req'd. (<3)	●	●	●	●	●	●	●	●	●	●	●	●	●	●	●	X	X	X	X	X
	Low resistance req'd. (3–10)	●	●	●	●	●	●	●	●	●	●	●	●	●	●	●	●	●	●	●	●
	Moderate resistance tolerable (10–30)	●	●	●	●	●	●	●	●	●	●	●	●	●	●	●	●	●	●	●	●
	High resistance tolerable (>30)	●	●	●	●	●	●	●	●	●	●	●	●	●	●	●	●	●	●	●	●
	Piggable	X	X	X	X	●	X	●	X	X	●	●	●	●	●	●	X	X	X	X	X
Fluid	Liquid (and two phase)	●	●	●	●	●	●	●	●	X	●	●	●	●	●	●	●	●	B	B	●
	Gas	●	●	●	●	●	●	●	●	X	C	●	●	●	●	●	●	●	●	●	●
	Steam	D	●	●	X	X	X	X	●	X	X	R	R	X	X	X	X	X	X	X	X
	Slurry	X	X	X	C	X	X	X	X	●	C	C	●	C	C	●	●	●	X	X	C
	Solids (powder etc.)	X	X	X	C	X	X	X	X	●	C	C	X	X	X	X	C	C	X	X	X
Fluid condition	Clean	●	●	●	●	●	●	●	●	●	●	●	●	●	●	●	●	●	●	●	●
	Dirty (abrasive)	E	E	E	E	●	X	X	X	E	●	●	●	X	X	X	E	E	E	E	●
	Large susp. solids	X	X	X	X	X	X	X	X	●	●	●	X	X	X	X	E	E	●	●	C
	Solidifying	J	J	J	X	X	X	X	X	J	J	C	J	X	X	C	C	X	C	C	C
	Viscous	●	●	C	●	●	C	C	X	●	●	●	●	●	●	●	●	●	●	●	●
	Corrosive	L	L	L	L	L	L	L	L	L	L	L	L	L	L	L	L	L	L	L	L
	Flammable	●	●	●	X	●	●	●	X	X	N	N	N	M	M	M	N	N	N	N	N
	Fouling scaling	X	X	X	X	●	X	X	X	X	●	●	●	X	X	X	●	●	G	G	C
	Searching	P	P	P	X	X	C	C	X	X	X	X	X	P	P	P	P	P	X	X	P

Notes on flow resistance
Valve flow resistances are presented as multiples of the resistance of a plain piece of pipe where this is equivalent to 1 (the figures are approximate).
Reduced port (venturi) gate and ball valves may have up to twice the flow resistance of full bore valves.

| Plug parallel | | | Diaphragm | | | | Globe | | | | Butterfly | | | | | |
Expanding plug	Eccentric plug	Lubricated	Full bore	Weir	Pinch	Iris	Straight	Angle	Oblique ('Y' type)	Needle	Concentric, metal disk/seat	Concentric, rubber lined	Concentric, other lining	Eccentric, metal disk/seat	Eccentric, rubber lined	Eccentric, soft seat
X	X	X	X	X	X	X	X	X	X	X	X	X	X	X	X	X
•	•	•	•	X	X	X	X	X	X	X	A	A	A	A	A	A
•	•	•	•	•	•	•	X	X	•	X	•	•	•	•	•	•
•	•	•	•	•	•	•	•	•	•	•	•	•	•	•	•	•
X	X	X	X	X	X	X	X	X	X	X	X	X	X	X	X	X
•	•	B	•	•	•	•	X	•	•	•	•	•	•	•	•	•
•	•	•	•	•	X	X	•	•	•	•	X	•	C	C	•	•
X	X	X	X	X	X	X	•	•	•	•	•	X	X	C	X	X
X	C	X	•	X	•	X	X	X	•	•	•	•	C	•	C	
X	X	X	•	X	•	•	X	X	X	X	C	C	C	C	C	C
•	•	•	•	•	•	•	•	•	•	•	•	•	•	•	•	•
F	F	F	•	•	•	•	G	G	G	G	H	H	H	I	I	H
X	X	•	C	X	C	X	X	X	X	X	X	X	X	X	X	X
X	X	C	X	X	X	X	X	X	X	X	X	X	X	X	X	X
•	•	•	•	X	•	X	G	G	G	G	•	•	•	•	•	•
L	L	X	K	K	K	L	L	L	L	L	L	L	L	L	L	L
M	M	X	X	X	X	•	•	•	•	X	X	X	•	X	X	M
X	X	X	X	X	X	X	X	X	X	X	•	X	X	•	X	X
C	C	X	K	K	K	X	P	P	P	P	X	X	X	X	X	P

Key

• Suitable

× Not suitable/not recommended

A Some butterfly valves may fall into the 'low resistance' category

B Depends on liquid. Unsuitable for use with solvents etc

C May be suitable: consult manufacturer

D Flexi-wedge more suitable for this service in large sizes

E May be used subject to nature of fluid. Sharp particles may be trapped in cavities and damage soft seats

F No information available, but unlikely to be suitable

G Not normally recommended

H Variable performance. Moderate service life

I Can perform well; depends on manufacturer

J Must be full bore. Steam jacket/trace heating required

K Must have secondary stem seal

L Satisfactory subject to appropriate choice of materials. Careful attention should be given to design of internal parts etc

M Fire-tested type required

N Fire-tested or fire-resistant gland required. Plug valve sleeves and linings may be resistant to fire but do not provide shut-off capability after destruction

P Use bellows sealed versions in smaller sizes where available. Helium leak test and double block and bleed for hydrogen service

R Only suitable if all plastic/rubber components eliminated

Table A2   Selection of isolation (block) valves – service conditions (*cont.*)

		Valve types		Wedge gate				Parallel gate					Ball metal seat			Ball soft seat			Plug taper				
		Valve description	Conditions	Solid wedge	Flexi-wedge	Split-wedge	Rubber lined	Conduit slab gate	Parallel double disk gate	Conduit split gate	Parallel slide (steam/feed)	Knife gate	Floating ball	Trunnion mounted ball	Eccentric ball	Floating ball	Trunnion mounted ball	Eccentric ball	Non-lubricated (sleeved)	Non-lubricated (lined)	Lubricated	Lubricated (balanced plug)	Lifting plug
Pressure		Vacuum		J	J	J	X	X	J	X	X	X	X	X	X	J	X	J	J	J	X	X	J
		Low ≤ CL. 150		●	●	●	●	●	●	●	B	●	●	●	●	●	●	●	●	●	●	●	●
		Med. CL. 300/600		●	●	●	●	●	●	●	●	X	●	●	●	●	●	●	●	●	●	●	●
		High CL. 900/2500		●	●	●	X	●	●	●	●	X	●	●	●	C	●	●	X	X	X	●	●
		Class 800		●	●	●	X	X	X	X	●	X	●	X	X	●	X	X	X	X	X	●	X
Temp.		Cryogenic < –50°C		●	●	●	X	X	X	X	X	X	X	X	X	●	●	X	X	X	X	X	X
		Med/low –50°C/200°C		●	●	●	A	●	●	●	●	●	●	●	●	A	A	A	●	●	A	A	●
		High 200°C/450°C		●	●	●	X	A	A	A	●	X	A	A	A	X	X	X	X	X	X	X	●
Size		≤ 1½" NS (40 DN)		●	X	X	X	●	X	X	●	X	●	X	●	●	X	●	●	●	●	●	●
		2"–8" NS (50–200 DN)		●	●	●	●	●	●	●	●	●	●	●	●	●	●	●	●	●	●	●	●
		10"–16" NS (250–400 DN)		●	●	●	●	●	●	●	●	●	X	●	F	X	●	F	F	F	●	●	F
		> 16" NS (400 DN)		●	●	●	●	●	●	●	●	●	X	●	F	X	●	F	X	X	●	●	F
Isolation		Bubble tight (gas)		H	H	H	●	X	H	H	X	X	X	X	H	G	G	●	●	●	X	X	H
		Drop tight (liquid)		H	H	H	●	X	H	H	X	H	X	X	●	G	●	●	●	●	I	I	●
		Very low leakage permitted		●	●	●	●	X	●	●	●	H	X	X	●	●	●	●	●	●	I	I	●
		Some leakage permitted (normal commercial std)		●	●	●	●	●	●	●	●	H	●	●	●	●	●	●	●	●	●	●	●
		Course isolation only		●	●	●	●	●	●	●	●	●	●	●	●	●	●	●	●	●	●	●	●

Notes

The degree of isolation quoted is that which is consistently and readily achievable and can be maintained in service.

It should be noted that most metal seated valves can be produced to achieve a greater degree of leak tightness than that shown in this table. This will require greater expenditure of time and effort, but the degree of isolation thus achieved is likely to be maintained for longer than that more readily available by use of soft seated valves.

Plug parallel			Diaphragm				Globe				Butterfly					
Expanding plug	Eccentric plug	Lubricated	Full bore	Weir	Pinch	Iris	Straight	Angle	Oblique ('Y' type)	Needle	Concentric, metal disk/seat	Concentric, rubber lined	Concentric, other lining	Eccentric, metal disk/seat	Eccentric, rubber lined	Eccentric, soft seat
J	J	X	X	X	X	X	J	J	J	X	X	K	X	X	K	K
●	●	●	●	●	●	●	●	●	●	●	●	●	●	●	●	●
●	X	●	X	X	X	X	●	●	●	●	●	●	●	●	●	●
●	X	X	X	X	X	X	●	●	●	●	X	X	X	X	X	X
X	X	X	X	X	X	X	●	●	●	●	X	X	X	X	X	X
X	X	X	X	X	X	X	●	●	X	●	X	X	●	●	X	●
A	●	A	A	A	A	●	●	●	●	●	●	A	A	●	A	A
X	X	X	X	X	X	X	●	●	●	●	●	X	X	●	X	X
X	●	X	●	●	●	X	●	●	●	●	X	X	X	X	X	X
●	●	●	●	●	●	●	●	●	●	E	●	●	●	●	●	●
F	●	●	F	F	●	●	●	●	●	X	●	●	●	●	●	●
F	●	●	X	X	X	X	X	X	X	X	●	●	●	●	●	●
●	●	X	●	●	X	X	H	H	H	H	X	●	X	X	●	●
●	●	I	●	●	●	X	H	H	H	H	X	●	X	X	●	●
●	●	I	●	●	●	X	●	●	●	●	X	●	●	●	●	●
●	●	●	●	●	●	X	●	●	●	●	X	●	●	●	●	●
●	●	●	●	●	●	●	●	●	●	●	●	●	●	●	●	●

Key
- ●   Suitable/available/achievable
- ×   Not suitable/available/achievable
- A   Temperature range may be limited by soft seats/seals/linings etc
- B   Relies on differential pressure for seal, poor sealing at very low pressures
- C   <2" NB only
- E   Not recommended in sizes larger than 2" NB
- F   Limited size range depending on pressure rating
- G   May not be achievable at low pressures
- H   Usually only achievable by soft seated valves
- I   Leak tightness depends on efficiency of sealant
- J   Soft seated valves preferred. Consult manufacturer re: seat finishing, cleaning. Consider bellows seals. Helium leak test recommended
- K   May be suitable: consult manufacturer

Table A3   Selection of isolation (block) valves – features

	Features	Solid wedge	Flexi-wedge	Split-wedge	Rubber lined	Conduit slab gate	Parallel double disk gate	Conduit split gate	Parallel slide (steam/feed)	Knife gate	Floating ball	Trunnion mounted ball	Eccentric ball	Floating ball	Trunnion mounted ball	Eccentric ball	Non-lubricated (sleeved)	Non-lubricated (lined)	Lubricated	Lubricated (balanced plug)	Lifting plug
	Valve types →	Wedge gate				Parallel gate					Ball metal seat			Ball soft seat			Plug taper				
Gland packing	Std. stuffing box	●	●	●	●	●	●	●	●	X	B	B	●	B	B	●	B	A	A	A	●
	'O' rings	B	B	B	B	B	B	B	A	X	●	●	X	●	●	X	A	A	●	●	A
	Polymer seals	A	A	A	A	●	●	●	A	X	●	●	●	●	●	●	●	●	B	B	B
	Fire tested	X	X	X	X	X	X	X	X	X	●	●	●	●	●	●	●	●	●	●	A
	Extended bonnet (LT)	●	B	A	X	A	A	A	A	X	A	A	A	●	●	A	X	X	X	X	X
	Bellows seal	C	X	X	X	X	A	A	A	X	X	X	X	C	C	X	X	X	X	X	X
Linings	Rubber	●	X	X	●	X	X	X	X	X	X	X	X	X	X	X	X	X	X	X	●
	PTFE	A	A	A	X	X	X	X	X	X	X	X	X	●	●	●	X	●	X	X	A
	Other polymers	B	A	A	X	B	B	B	X	X	X	X	X	B	B	B	X	B	X	X	A
	Glass	X	X	X	X	X	X	X	X	X	X	X	X	X	X	X	X	X	X	X	X

Expanding plug	Eccentric plug	Lubricated	Full bore	Weir	Pinch	Iris	Straight	Angle	Oblique ('Y' type)	Needle	Concentric, metal disk/seat	Concentric, rubber lined	Concentric, other lining	Eccentric, metal disk/seat	Eccentric, rubber lined	Eccentric, soft seat
●	●	A	X	X	X	X	●	●	●	●	●	A	A	●	A	●
A	A	A	B	B	B	X	B	B	A	●	●	●	●	●	●	●
A	●	●	X	X	X	X	C	C	C	C	A	A	A	●	●	●
●	●	A	X	X	X	X	X	X	X	X	●	X	X	●	X	●
X	X	X	X	X	X	X	●	●	X	●	●	X	A	●	X	●
X	X	X	X	X	X	X	C	B	B	C	X	X	X	X	X	X
X	●	X	●	●	●	X	X	X	X	X	X	●	X	X	●	X
D	A	X	A	●	A	X	D	D	D	D	X	X	●	X	X	●
D	●	X	A	●	A	X	B	D	B	D	X	X	●	X	X	●
X	A	X	●	●	X	X	X	X	X	X	X	X	X	X	X	X

Key

- ● Available
- × Not available/not applicable
- A Not normally available
- B Limited availability
- C Available sizes may be limited
- D Not known

Table A4　Selection of isolation (block) valves – available materials

	Material	Solid wedge	Flexi-wedge	Split-wedge	Rubber lined	Conduit slab gate	Parallel double disk gate	Conduit split gate	Parallel slide (steam/feed)	Knife gate	Floating ball	Trunnion mounted ball	Eccentric ball	Floating ball	Trunnion mounted ball	Eccentric ball	Non-lubricated (sleeved)	Non-lubricated (lined)	Lubricated	Lubricated (balanced plug)	Lifting plug
		Wedge gate				Parallel gate					Ball metal seat			Ball soft seat			Plug taper				
Body, bonnet etc.	Carbon steel	●	●	●	●	●	●	●	●	●	●	●	●	●	●	●	●	●	●	●	●
	Chrome moly steel	●	●	●	X	A	●	A	●	X	A	A	A	X	X	X	X	X	A	A	A
	Aust. stainless steel	●	●	●	X	●	B	B	B	●	●	●	●	●	●	●	X	●	●	●	●
	Nickel alloys	B	B	A	X	●	B	B	B	●	●	●	●	●	●	●	●	X	A	A	B
	Aluminium bronze	B	B	A	X	A	A	A	X	A	X	X	A	B	B	A	A	A	A	A	A
	Bronze/gunmetal	●	●	A	X	A	A	A	●	A	A	A	A	●	●	A	A	A	●	A	A
	Aluminium	B	A	A	X	A	A	A	A	A	A	A	A	●	A	A	A	A	A	A	A
	Cast iron	●	●	A	X	A	A	A	●	B	A	A	A	●	A	A	●	●	●	●	●
	PVC	B	X	X	X	X	X	X	X	X	X	X	X	●	X	X	X	X	X	X	X
	Cupro-nickel	●	●	A	X	A	A	A	A	A	A	A	A	●	●	A	●	●	A	A	A
	Glass	X	X	X	X	X	X	X	X	X	X	X	X	●	A	X	X	X	X	X	X
Trim	13% Cr (410) SS	●	●	●	●	A	A	A	A	A	X	X	X	●	●	A	A	A	A	●	●
	13% Cr, Ni faced	A	A	A	A	A	A	A	A	A	A	A	A	A	A	A	A	A	A	A	A
	13% Cr, hard faced	●	●	●	X	●	●	●	●	A	A	A	A	A	A	A	A	A	A	A	A
	Austenitic SS	A	A	A	A	A	●	●	A	●	A	A	A	●	●	●	●	A	●	●	●
	Aust. SS hard faced	●	●	●	A	●	●	●	●	B	●	●	●	B	B	●	A	A	A	A	A
	CS hard faced	A	●	A	X	●	●	●	●	A	A	●	A	A	A	X	X	X	X	X	X
	Monel/inconel	●	●	A	A	A	A	A	●	●	A	A	●	●	A	●	●	A	A	A	A
	Chrome plated	A	A	A	A	B	B	B	A	B	A	●	A	●	●	A	A	A	A	A	B
	ENP	A	A	A	A	●	B	B	A	B	A	●	A	●	●	A	X	X	●	A	●
	Cast iron	●	A	A	●	A	A	A	A	A	A	A	A	A	A	●	●	●	●	●	●
	Bronze	●	A	A	A	A	A	A	●	A	A	A	A	●	B	A	A	A	C	A	A

Plug parallel			Diaphragm				Globe				Butterfly					
Expanding plug	Eccentric plug	Lubricated	Full bore	Weir	Pinch	Iris	Straight	Angle	Oblique ('Y' type)	Needle	Concentric, metal disk/seat	Concentric, rubber lined	Concentric, other lining	Eccentric, metal disk/seat	Eccentric, rubber lined	Eccentric, soft seat
●	●	●	B	B	●	●	●	●	●	●	●	●	●	●	●	●
A	A	A	X	X	X	X	●	●	●	●	B	X	X	B	X	X
●	●	●	B	●	X	A	●	●	●	●	●	X	X	●	X	●
B	●	B	A	A	A	A	B	B	B	B	●	A	A	B	X	A
A	A	A	A	A	A	A	C	C	C	C	B	X	X	B	X	B
A	●	A	●	●	A	A	●	●	●	●	B	X	X	B	X	X
A	●	A	A	A	A	A	C	C	C	C	B	X	X	B	X	X
●	●	●	●	●	●	D	●	●	●	●	●	●	●	●	●	●
X	X	B	D	●	D	D	D	●	D	D	X	X	●	X	X	●
A	A	A	A	A	A	A	C	C	C	C	B	X	X	B	X	X
X	X	●	X	X	X	X	X	●	●	X	X	X	X	X	X	X
A	A	●	X	X	X	X	●	●	●	●	A	A	A	A	A	A
A	A	A	X	X	X	X	B	B	B	B	A	A	A	A	A	A
A	A	A	X	X	X	X	B	B	B	B	A	A	A	●	A	A
●	●	●	X	X	X	X	●	●	●	●	●	●	●	●	●	●
A	A	A	X	X	X	X	●	●	●	●	A	A	A	●	A	A
X	X	X	X	X	X	X	B	B	X	X	X	X	X	X	X	X
A	●	A	X	X	X	X	B	B	B	B	●	B	B	●	B	B
●	A	B	X	X	X	X	A	A	A	B	B	B	B	●	B	B
●	●	●	X	X	X	X	A	A	A	●	A	A	●	A	A	●
●	●	●	X	X	X	X	●	●	A	A	●	●	●	●	●	●
B	●	B	X	X	X	X	●	●	●	●	B	B	●	B	B	●

Key
● Available
× Not available/not applicable
A Not normally available
B Limited availability
C Available sizes may be limited
D Not known

Examples of carbon and alloy steels available:
Carbon steel
Carbon steel (low temperature)
Carbon/molybdenum steel
$1\frac{1}{4}$Cr–$\frac{1}{2}$Mo
$2\frac{1}{4}$Cr–1Mo
3Cr–1Mo
5Cr–$\frac{1}{2}$Mo
9Cr–1Mo
2Ni
$3\frac{1}{2}$Ni

Examples of corrosion-resistant alloys available:

Alloy	Description
18Cr–10Ni	304 austenitic stainless steel
18Cr–10Ni–2Mo	316 austenitic stainless steel
Ni–Cr–Fe	Inconel
Ni–Cu	Monel
Ni–Mo	Hastelloy B
Ni–Mo–Cr	Hastelloy C

Other materials (e.g., duplex, SS, titanium) may be available to special order.
For acid and other non-fire hazardous, non-hydrocarbon services, plastic or rubber lined valves may be considered.
Note that, in some cases, valve trim may consist of a combination of several materials. For information on applications of materials see supplement.

Table A5   Selection of isolation (block) valves – achievable leakage rates

Valve types / Leakage rate	Wedge gate				Parallel gate					Ball metal seat			Ball soft seat			Plug taper				
	Solid wedge	Flexi-wedge	Split-wedge	Rubber lined	Conduit slab gate	Parallel double disk gate	Conduit split gate	Parallel slide (steam/feed)	Knife gate	Floating ball	Trunnion mounted ball	Eccentric ball	Floating ball	Trunnion mounted ball	Eccentric ball	Non-lubricated (sleeved)	Non-lubricated (lined)	Lubricated	Lubricated (balanced plug)	Lifting plug
Liquid																				
ISO 5208 rate 3	B	B	B	●	X	B	B	X	B	X	X	●	A	●	●	●	●	F	F	●
ISO 5208 rate 2	●	●	●	●	X	●	●	●	B	X	X	●	●	●	●	●	●	F	F	●
ISO 5208 rate 1	●	●	●	●	●	●	●	●	B	●	●	●	●	●	●	●	●	●	●	●
API 598	C	C	C	●	C	C	C	C	B	D	D	D	A	●	●	●	●	●	●	●
MSS–SP–61	●	●	●	●	●	●	●	●	B	●	●	●	●	●	●	●	●	●	●	●
BS 6755 Pt 1 rate C	●	●	●	●	●	●	●	●	B	X	●	●	●	●	●	●	●	F	F	●
Gas																				
ISO 5208 rate 3	B	B	B	●	X	B	B	X	X	X	X	B	A	A	●	●	●	X	X	B
ISO 5208 rate 2	B	B	B	●	X	B	B	X	X	X	X	●	A	●	●	●	●	X	X	B
ISO 5208 rate 1	●	●	●	●	●	●	●	●	X	X	●	●	●	●	●	●	●	●	●	●
API 598	B	B	B	●	X	B	B	X	X	D	D	D	A	●	●	●	●	X	X	B
MSS–SP–61	●	●	●	●	●	●	●	●	X	X	●	●	●	●	●	●	●	●	●	●
BS 6755 Pt 1 rate C	●	●	●	●	●	●	●	●	X	X	A	●	A	●	●	●	●	F	F	●

Leakage rates (ml/min/mm DN)

Test medium	ISO 5208 Rate 3 BS 6755 Pt 1 Rate A	ISO 5208 Rate 2 BS 6755 Pt 1 Rate B	BS 6755 Pt 1 Rate C	ISO 5208 Rate 1 BS 6755 Pt 1 Rate D MSS–SP–61
Liquid	No visually detectable leakage for the duration of the test	0.0006	0.0018	0.006
Gas	No visually detectable leakage for the duration of the test	0.018	0.18	1.8

API 598 leakage rates are not directly comparable, but the following may be used as a guide:
- for soft seated valves, leakage rate corresponds to ISO 5208 rate 3;
- for metal seated gate, globe, and plug valves liquid leakage rate approximately corresponds to ISO 5208 rate 3 for valves ≤2″ NB, ISO 5208 rate 1 for sizes between 2″, and 12″ NB and is between BS 6755 rate C and ISO 5208 rate 2 for valves ≥14″ NB. Gas leakage is approximately half that of ISO 5208 rate 2;
- for metal seated ball and butterfly valves leakage rates are by agreement with the purchaser.

Note: a leakage rate of one drop of liquid corresponds to 0.0625 ml
a leakage rate of one bubble (from $\frac{1}{4}$″ dia. tube) corresponds to 0.15 ml (ANSI B16 104)

| Plug parallel | | | Diaphragm | | | | Globe | | | | Butterfly | | | | | |
Expanding plug	Eccentric plug	Lubricated	Full bore	Weir	Pinch	Iris	Straight	Angle	Oblique ('Y' type)	Needle	Concentric, metal disk/seat	Concentric, rubber lined	Concentric, other lining	Eccentric, metal disk/seat	Eccentric, rubber lined	Eccentric, soft seat
●	●	F	●	●	●	E	B	B	B	B	E	●	×	×	●	●
●	●	F	●	●	●	E	●	●	●	●	E	●	●	●	●	●
●	●	●	●	●	●	E	●	●	●	●	E	●	●	●	●	●
●	●	C	●	●	●	E	C	C	C	C	E	●	×	D	●	●
●	●	●	●	●	●	E	●	●	●	●	E	●	●	●	●	●
●	●	F	●	●	●	E	●	●	●	●	E	●	●	●	●	●
●	●	×	●	●	×	E	B	B	B	B	E	●	×	×	●	●
●	●	×	●	●	×	E	B	B	B	B	E	●	×	×	●	●
●	●	●	●	●	●	E	●	●	●	●	E	●	●	●	●	●
●	●	×	●	●	×	E	B	B	B	B	E	●	×	D	●	●
●	●	●	●	●	●	E	●	●	●	●	E	●	●	●	●	●
●	●	F	●	●	×	E	●	●	●	●	E	●	×	×	●	●

Key

●   Consistently achievable

×   Not normally achievable

Note: By expenditure of sufficient time and effort, it is usually possible to achieve a high degree of seat leak tightness with most types of valve. In general, however, it is not good practice to specify leakage rates which are unnecessarily stringent compared to actual process requirements

A   May not be achievable at low pressures

B   Usually only achievable by soft seated valves

C   Valves 2″ NB and less may need to be soft sealed

D   Leak rates are determined by agreement with the purchaser

E   Valve not intended for tight shut-off on liquid or gas service

F   Leak tightness depends on efficiency of sealant

Table A6   Selection of check and diverter valves – service conditions

		Check valves																		
Valve types		Lift								Swing						Other			
	Valve description / Conditions	Ball	Disk	Piston (axial)	Ball (spring loaded)	Piston (spring loaded)	Plate (spring loaded)	Lift type with dash pot	Lift type—angle pattern	Swing with long arm	Swing—standard pattern	Wafer pattern	Tilting disk	Swing—spring loaded	Dual-disk—spring loaded	Foot valves	Screw-down stop & check	Axial flow anti-slam	Diaphragm
Resistance to flow	Low resistance required (< 30)	X	X	X	X	X	X	X	X	●	●	X	●	●	●	X	X	●	X
	Moderate resistance tolerable (30–50)	X	X	●	X	●	X	X	●	●	●	X	●	●	●	●	X	●	●
	High resistance tolerable (> 50)	●	●	●	●	●	●	●	●	●	●	●	●	●	●	●	●	●	●
	Piggable	X	X	X	X	X	X	X	X	A	A	X	X	A	X	X	X	X	X
Fluid	Liquid (& two phase)	●	●	●	●	●	●	●	●	●	●	●	●	●	●	●	●	●	●
	Gas	●	●	●	●	●	●	●	●	●	●	●	●	●	●	X	●	●	X
	Steam	●	●	●	●	●	●	●	●	●	●	●	●	●	●	X	●	●	X
	Slurry	X	X	X	X	X	X	X	X	●	●	●	X	X	X	X	X	X	●
	Solids (powder etc.)	X	X	X	X	X	X	X	X	X	X	X	X	X	X	X	X	X	●
Fluid condition	Clean	●	●	●	●	●	●	●	●	●	●	●	●	●	●	●	●	●	●
	Dirty (abrasive)	B	B	X	B	X	X	X	B	B	B	B	B	B	●	B	X	X	●
	Large susp. solids	X	X	X	X	X	X	X	X	C	C	X	X	C	X	X	X	X	D
	Solidifying	F	X	X	F	X	X	X	F	F	F	F	X	F	F	X	X	X	X
	Viscous	●	F	F	●	F	X	F	F	F	F	F	F	F	F	F	F	F	F
	Corrosive	G	G	G	G	G	G	G	G	G	G	G	G	G	G	G	G	G	G
	Flammable	H	H	H	H	H	H	H	H	H	H	H	H	H	H	X	H	H	X
	Fouling/scaling	E	E	E	E	E	E	E	E	E	E	E	E	E	E	E	E	E	E
	Searching	I	I	I	I	I	X	I	I	I	I	I	I	I	I	X	I	●	X

Diverter valves					
3 & 4 way taper plug (sleeved)	3 & 4 way taper plug (lubricated)	4 way lifting plug (taper)	4 way expanding plug (parallel)	3 way ball (metal seated)	3 way globe valve
●	●	●	●	●	X
●	●	●	●	●	X
●	●	●	●	●	●
X	X	X	X	X	X
●	J	●	●	●	●
●	●	●	●	X	●
X	X	X	X	X	●
●	X	C	X	●	X
C	X	X	X	●	X
●	●	●	●	●	●
K	K	C	C	●	P
C	●	C	X	●	X
X	C	C	X	●	X
●	●	●	●	●	P
L	L	L	L	L	L
M	M	M	M	M	M
●	X	C	X	●	X
N	X	N	N	X	N

Key

● Suitable

× Not suitable/not recommended

A Can be pigged provided valves specially designed

B Hard particles may prevent complete closure of valve. Valves not in fully open position may suffer wear of hinge pins etc

C Consult manufacturer. Seats may suffer damage and sealing ability may be impaired

D Depends on type

E Check valves not really suitable for this service and likely to give poor performance

F Consult manufacturer. Steam jacketing/trace heating may be required. Valves without spring assistance likely to be sluggish in operation

G Satisfactory subject to appropriate choice of materials. Careful attention should be given to design of springs and other internal parts

H Valves incorporating soft sealing components may need to be fire tested

I Some leakage likely, even with soft seat materials. Helium leak test for hydrogen service

J Depends on liquid. Unsuitable for use with solvents etc

K May be used subject to nature of fluid. Sharp particles may be trapped in cavities and damage soft seats

L Satisfactory subject to appropriate choice of materials. Careful attention should be given to design of internal parts etc

M Fire tested or fire-resistant gland required

Notes on flow resistance

Valve flow resistances are presented as multiples of the resistance of a plain piece of pipe where this is equivalent to 1 (the figures are approximate).

There are no designs of check valve offering the very low flow resistance of ball and gate valves. If minimum flow resistance is the dominating criterion of selection, the axial flow, anti-slam valve should be chosen.

Table A7 Selection of check and diverter valves – service conditions and sizes

	Valve types →	Check valves — Lift								Check valves — Swing						Check valves — Other				Diverter valves					
Group	Conditions	Ball	Disk	Piston (axial)	Ball (spring loaded)	Piston (spring loaded)	Plate (spring loaded)	Lift type with dash pot	Lift type – angle pattern	Swing with long arm	Swing – standard pattern	Wafer pattern	Tilting disk	Swing – spring loaded	Dual-disk – spring loaded	Foot valves	Screw-down stop & check	Axial flow anti-slam	Diaphragm	3 & 4 way taper plug (sleeved)	3 & 4 way taper plug (lubricated)	4 way lifting plug (taper)	4 way expanding plug (parallel)	3 way ball (metal seated)	3 way globe valve
Pressure	Vacuum	X	X	X	B	B	B	X	X	X	X	X	X	X	B	X	X	B	B	X	X	B	B	X	B
Pressure	Low ≤ CL. 150	●	●	●	●	●	●	●	●	●	●	●	●	●	●	●	●	●	●	●	●	●	●	●	●
Pressure	Med. CL. 300/600	●	●	●	●	●	●	●	●	●	●	●	●	●	●	X	●	●	X	G	G	●	X	●	●
Pressure	High CL. 900–2500	C	D	●	C	C	●	C	C	●	●	D	●	X	●	X	C	●	X	X	X	G	X	D	●
Pressure	Class 800	●	X	●	●	●	X	●	●	E	●	X	E	X	X	X	X	X	X	X	X	X	X	X	X
Temp.	Cryogenic < –50°C	F	F	F	X	X	X	F	F	X	F	X	F	X	X	X	X	●	X	X	X	X	X	E	●
Temp.	Med/low –50°C/200°C	●	●	●	●	●	●	●	●	●	●	●	●	●	●	●	●	●	A	●	●	●	●	●	●
Temp.	High 200°C to 450°C	●	●	●	●	●	●	●	●	●	●	X	●	●	●	●	●	X	X	X	X	X	X	●	●
Size	≤ 1½" NS (40 DN)	●	●	●	●	●	●	●	●	X	X	X	X	X	X	X	●	X	X	●	●	X	X	E	●
Size	2"–8" NS (50–200 DN)	●	D	●	D	●	●	●	●	●	●	●	●	●	●	●	●	●	●	●	●	●	●	●	●
Size	10"–16" NS (250–400 DN)	G	X	E	X	E	G	D	D	●	●	●	●	●	●	●	E	●	●	X	X	G	●	E	D
Size	> 16" NS (400 DN)	X	X	X	X	X	X	X	X	●	●	●	●	●	G	●	X	●	D	X	X	X	●	X	E
Isolation	Bubble tight (gas)	H	H	H	H	H	H	H	H	X	X	X	X	X	H	X	H	H	X	●	X	●	●	X	A
Isolation	Drop tight (liquid)	H	H	H	H	H	H	H	H	X	X	X	X	X	H	X	H	H	●	●	K	●	●	X	H
Isolation	Very low leakage permitted	I	I	I	I	I	I	I	I	I	I	I	I	I	I	●	I	●	●	●	K	●	●	X	●
Isolation	Some leakage permitted (normal commercial std)	●	●	●	●	●	●	●	●	●	●	●	●	●	●	●	●	●	●	●	●	●	●	●	●
Isolation	Course isolation only	●	●	●	●	●	●	●	●	●	●	●	●	●	●	●	●	●	●	●	●	●	●	●	●
Operation	Pulsating flow	X	X	X	●	●	J	●	●	X	X	X	X	X	X	●	●	●	●						
Operation	Unstable flow	X	X	X	X	X	X	X	X	X	X	X	●	X	X	X	X	●	X						
Operation	Sudden flow	●	X	X	●	●	●	●	●	X	●	X	●	●	●	●	●	●	●						
Operation	Sudden flow reversal	X	X	X	●	●	●	●	X	X	X	X	●	X	●		X	●	●						

Key

● Suitable/available/achievable

x Not suitable/available/achievable

A Temperature range may be limited by soft seats/seals/linings etc

B Soft seated valves only. Few check valves are suitable for sealing against vacuum. Helium leak test recommended

C Small sizes only

D Limited availability

E Not normally available

F Consult manufacturer; special design required

G Limited pressure rating

H Usually only achievable by soft seated valves

I May not be achievable at low pressures except by soft seated valves. Even at higher pressures, metal seated valves are likely to require special effort to achieve this

J Compressor discharge valves only

K Leak tightness depends on efficiency of sealant

Table A8 Selection of check and diverter valves – available materials

Valve types (Materials)	Lift: Ball	Disk	Piston (axial)	Ball (spring loaded)	Piston (spring loaded)	Plate (spring loaded)	Lift type with dash pot	Lift type – angle pattern	Swing: Swing with long arm	Swing – standard pattern	Wafer pattern	Tilting disk	Swing – spring loaded	Dual-disk – spring loaded	Other: Foot valves	Screw-down stop & check	Axial flow anti-slam	Diaphragm	Diverter: 3 & 4 way taper plug (sleeved)	3 & 4 way taper plug (lubricated)	4 way lifting plug (taper)	4 way expanding plug (parallel)	3 way ball (metal seated)	3 way globe valve	
Body, bonnet etc.																									
Carbon steel	●	●	●	●	●	●	●	●	●	●	●	●	●	●	●	●	●	B	●	●	●	●	B	●	
Chrome moly steel	A	●	A	A	●	●	●	●	A	●	A	●	A	A	X	●	X	X	X	X	X	X	B	●	
Aust. stainless steel	●	●	●	A	●	●	●	●	●	●	●	●	A	●	●	●	●	●	●	●	A	A	●	●	
Nickel alloys	X	B	B	X	A	A	A	A	A	B	A	A	X	B	X	X	B	X	●	A	A	A	A	●	B
Aluminium bronze	X	A	A	X	A	X	A	A	X	B	●	X	X	●	X	X	●	X	A	A	X	X	A	B	
Bronze/gunmetal	X	●	●	X	A	A	●	●	X	●	A	A	X	B	●	●	X	●	A	X	X	X	X	●	
Aluminium	C	C	C	C	C	C	C	C	C	C	C	C	C	C	C	C	C	C	A	X	X	X	X	A	
Cast iron	X	●	A	X	X	●	X	X	B	●	●	X	X	●	●	●	X	●	●	●	X	X	X	●	
PVC	●	X	X	X	X	X	X	●	X	X	X	X	X	X	●	X	X	X	X	X	X	X	X	C	
Cupro-nickel	X	C	X	X	C	X	C	C	X	B	X	X	X	B	X	X	X	X	●	A	A	A	A	B	
Glass	●	X	X	X	X	X	X	X	X	X	X	X	X	X	X	X	X	X	X	X	X	X	X	C	
Trim																									
13% Cr (410) SS	●	●	●	●	X	X	●	●	●	●	●	X	X	●	X	X	X	X							
Aust. stainless steel	●	●	●	●	●	●	●	●	●	●	●	●	A	●	●	●	●	●							
Hard faced	●	●	●	●	●	B	●	●	●	●	●	●	●	A	X	●	A	X							
Monel/inconel	X	B	B	X	●	X	A	A	A	B	A	A	X	B	X	B	B	X							
Aluminium bronze	X	A	A	X	A	X	A	A	X	B	●	X	X	●	X	X	●	X							
Bronze/gunmetal	X	●	●	X	A	X	●	●	X	●	A	A	X	B	●	●	X	X							
Linings																									
Rubber	●	X	X	X	X	X	X	X	X	X	●	X	X	●	X	X	X	●							
PTFE	●	●	●	X	X	X	X	X	X	X	●	X	X	X	X	X	X	X							
Other polymers	X	●	●	X	X	X	X	X	X	X	●	X	X	●	X	X	●	X							
Glass	●	X	X	X	X	X	X	X	X	X	X	X	X	X	X	X	X	X							

Key
● Available
× Not available/not applicable
A Not normally available
B Limited availability
C Not known

EXAMPLE 1: SELECTION OF AN ISOLATING VALVE FROM
TABLES A1 AND A2

- Application: Fluid   – clean acid, liable to cause scaling
                Temperature   – 150°C
                Pressure   – 70 bar rating
                Size   – DN 150
                Flow resistance – low
                Seat tightness – drop tight
- *Step 1* Consider flow resistance
  Inspect Table A1 for ability to provide low flow resistance.
  (All gate, ball, and plug valves are suitable plus full bore diaphragm
  valve, and possibly some butterfly.)
- *Step 2* Consider type of fluid
  Inspect valves from step (1) for suitability for liquid service.
  (Knife gate valves are excluded and lubricated plug valves are doubt-
  ful.)
- *Step 3* Consider fluid condition
  Fluid is clean, corrosive, and scaling.
  (Valves suitable for scaling are slab-gate, metal seated ball, sleeved
  plug valve, and metal seated butterfly.)
- *Step 4* Consider pressure
  Pressure is medium (Cl 600 Rating for carbon steel).
  (All remaining valve types suitable.)
- *Step 5* Consider temperature
  Temperature is medium/low.
  (All remaining valve types suitable.)
- *Step 6* Consider size
  (All remaining valve types suitable.)
- *Step 7* Consider seat leak tightness
  (The only one of the remaining valve types offering drop tight sealing
  according to the table is the sleeved plug valve. However, it should be
  noted that this is on the basis of long-term performance of the stan-
  dard product and that both metal seated butterfly and slab gate valves
  can be made to give comparable seat tightness.)
- *Step 8* Inspect Table A3 to check if the desired features are available
  with the selected valve type. If not, some compromise may have to be

made, either on features or by selecting (e.g.) an alternative valve type
with inferior sealing ability.
- *Step 9*   Inspect Table A4 to see if the chosen valve is available in the
required materials.

EXAMPLE 2: SELECTION OF A CHECK VALVE FROM TABLES A6 AND A7

- Application: Fluid   – clean, natural gas, valve at outlet
                                   of compressor on pipeline
               Temperature   – 60°C
               Pressure   – 150 bar
               Size   – DN 600
               Flow resistance – low; unstable flow conditions ex-
                                   pected occasionally
               Seat tightness   – very low leakage permitted
- *Step 1*   Consider flow resistance
  Inspect Table A6 for ability to provide low flow resistance.
  (All swing check types except wafer are suitable plus axial flow, anti-
  slam type.)
- *Step 2*   Consider type of fluid
  Inspect valves from step 1 for suitability to handle gas.
  (All are suitable.)
- *Step 3*   Consider fluid condition
  Fluid is clean.
  (All valves are suitable.)
- *Step 4*   Consider pressure
  Pressure is high (CL 1500 Rating), inspect Table A7 for suitability.
  (All except spring loaded swing check are suitable.)
- *Step 5*   Consider temperature
  Temperature is medium/low.
  (All remaining valves are suitable.)
- *Step 6*   Consider size limitations
  Valve is large (DN 600)
  (Dual disk, spring loaded check is not normally available in this size/
  pressure rating.)
- *Step 7*   Consider isolation capability

Only very low leakage is permissible.

(Most types will require soft seats to achieve this.)

– *Step 8*   Consider operation

Flow may be unstable.

(Of remaining valves, only tilting disk and axial flow, anti-slam type are suitable.)

Table A8 can be inspected for available materials etc., then decision made based on economics and further study of the detailed text.

APPENDIX B

Common Abbreviations

ABS	Acrylonitrile butadiene styrene
AFNOR	French Standards Institute
ANSI	American National Standards Institute
API	American Petroleum Institute
ASME	American Society of Mechanical Engineers
ASTM	American Society of Testing and Materials
BHRA	British Hydromechanics Research Group
BS	British Standard
BSI	British Standards Institute
CAF	Compressed asbestos fibre
CEN	European Standards Association
CFH	Cubic feet per hour
DN	Nominal diameter
DIN	German Standards Institute
EEMUA	Engineering Equipment and Materials Users Association
EPDM	Ethylene propylene (diene modified) rubber
EN	Euronorm (European standard)
FKM	Fluorinated rubber (viton)
GPDM	Ethylene propylene
GPM	Gallons per minute
HNBR	Hydrogenated nitrile rubber
HRC	Hardness, Rockwell, scale C7
IP	Institute of Petroleum
ISA	Instrument Society of America
ISO	International Standards Organisation
LNG	Liquefied natural gas
LME	Liquid metal embrittlement

LPG	Liquefied petroleum gas
MSS	Manufacturers Standardisation Society of the Valves and Fittings Industry (USA)
NACE	National Association of Corrosion Engineers
NB	Nominal bore
NBR	Nitrile rubber
NGL	Natural gas liquids
NPS	Nominal pipe size
PEEK	Polyetheretherketone
PP	Polypropylene
PTFE	Polytetrafluorethylene
PTFCE	Polytetrachlorofluoroethylene
PVC	Poly vinyl chloride
PVDF	Polyvinylidene fluoride
UPVC	Unplasticized poly vinyl chloride

APPENDIX C

List of Standards

The following list consists of currently available standards relating to isolation and check valves, some of which are referenced in the text. At the time of writing there is a great deal of activity by CEN to produce a range of European valve standards but few have been issued yet. Several API standards are being converted into ISO documents.

INTERNATIONAL DOCUMENTS

ISO 5208	Industrial valves–pressure testing for valves
ISO 10497	Testing of valves–fire type testing requirements

AMERICAN DOCUMENTS

ANSI B16.1	Cast iron pipe flanges and flanged fittings, classes 25, 125, 250, and 800
ANSI B16.5	Pipe flanges and flanged fittings
ANSI B16.10	Face to face and end to end dimensions of valves
ANSI B16.34	Valves – flanged and butt-welding end
API 6A	Specification for well head and Christmas tree equipment
API 6D	Specification for pipeline valves (steel gate, plug, ball, and check valves)
API 6FA	Specification for fire test for valves
API 14D	Well head surface safety valves
API 17D	Specification for sub-sea well head and Christmas tree equipment
API 594	Wafer-type check valves
API 597	Steel venturi gate valves, flanged, and butt-welding ends

API 599	Steel plug valves, flanged, and butt-welding ends
API 600	Steel gate valves, flanged, and butt-welding ends
API 602	Compact steel gate valves
API 603	Class 150, cast corrosion resistant, flanged end gate valves
API 606	Compact carbon steel gate valves (extended body)
API 607	Fire test for soft seated quarter turn valves
API 609	Butterfly valves, lug type and wafer type
API 941	Steels for hydrogen service at elevated temperatures and pressures in petroleum refineries and petrochemical plants
NACE Std MR–01–75	Material requirements – sulphide stress cracking
MSS SP 61	Pressure testing of steel valves
MSS SP 67	Butterfly valves
MSS SP 81	Stainless steel, bonnetless, flanged, wafer, knife gate valves

UNITED KINGDOM DOCUMENTS

BS 759 Pt.1	Specification for valve mountings and fittings
BS 1212	Specification for float operated valves
BS 1414	Steel wedge gate valves (flanged and butt-welding ends) for the petroleum, petrochemical, and allied industries
BS 1868	Steel check valves (flanged and butt-welding ends) for the petroleum, petrochemical and allied industries
BS 1873	Steel globe and globe stop and check valves (flanged and butt-welding ends) for the petroleum, petrochemical, and allied industries
BS 5150	Cast iron wedge and double disk gate valves for general purposes

BS 5146	Inspection and testing of valves
BS 5153	Cast iron check valves for general purposes
BS 5155	Specification for butterfly valves
BS 5156	Screw down diaphragm valves for general purposes
BS 5157	Steel gate (parallel slide) valves for general purposes
BS 5158	Cast iron and carbon steel plug valves for general purposes
BS 5163	Double flanged cast iron wedge gate valves for water works purposes
BS 5351	Steel ball valves for the petroleum, petrochemical, and allied industries
BS 5352	Specification for steel, wedge gate, globe and check valves 50 mm and smaller for the petroleum, petrochemical, and allied industries
BS 5353	Specification for plug valves
BS 6364	Specification for valves for cryogenic service
BS 6755 Pt. 1	Specification for production pressure testing requirements
BS 6755 Pt. 2	Specification for fire type-testing requirements
EEMUA Publ. No. 167	Specification for quality levels for carbon steel valve castings
EEMUA Publ. No 170 etc.	Specification for production testing of valves pts 1–4

APPENDIX D

Glossary of Valve Terminology

This section lists, in alphabetical order, many common terms used in valve specification and description.

Angle valve	A globe valve design with valve ends at right angles to each other. Normally with the inlet in the vertical plane and the outlet in the horizontal plane.
Abrasion	Damage to valve trim caused by hard particulate in the process fluid.
Actuator	A powered valve operator used to open or close a valve; energized by pneumatic, electric, or hydraulic power sources.
Air/vacuum relief valve	See pressure/vacuum safety valve.
Anti-blowout stem	A valve stem with a shoulder, positively retained by the body or bonnet. Typically a requirement for ball and butterfly valves.
Anti-static device	A device providing electric continuity between the valve body and internal components to prevent ignition of flammable fluids.
Automatic control valve	A valve automatically regulating the flow, pressure, or temperature of a fluid in response to a process signal from a sensing element.
Axial piston valve	A valve of the globe type having a piston-shaped closure with its polar axis in the line of flow. Generally used as a check valve, but with modification also used for control and block valve functions (also known as a nozzle valve, inline globe, or piston valve).

Back pressure	The pressure at the outlet, downstream of a safety valve. Back pressure is expressed as a percentage of the set pressure.
Back pressure controller	An automatic control valve with an internal sensing element which maintains a constant pressure in the upstream pipework.
Back seat	A face on the valve stem which seats on the underside of the gland stuffing box and provides a metal-to-metal seat on a full open valve. Generally provided on gate and globe valves.
Balanced safety valve	A safety valve in which back pressure exerts an equal force on the disk in both opening and closing directions (generally by means of a bellows).
Ball	See Closure member.
Ball check valve	A lift check valve having a free or spring loaded ball closing on a spherical seat (see also Check valve).
Ball valve	A quarter turn, rotary action valve with spherical closure and seats. Also called a ball plug valve for some design features.
Bellows	A convoluted, cylindrical component, usually metal, providing axial flexibility and pressure containment. Used as a gland seal (in bellows sealed valves) or for equalizing back pressure in balanced safety valves. Specifically strong designs may be used for mechanical loading seats in some designs of ball and gate valves for high temperature or dirty services.
Bellows sealed valve	A valve having a bellows sealed gland.
Bi-directional valve	A valve designed to seal against flow or pressure from either direction.
Block valve	A general term for valves used to shut-off flow and pressure. Other terms used are isolation valve, shut-off valve, and stop valve.

Blow down	A term, which when applied to safety valves, means the pressure difference between set pressure and reseating pressure, usually expressed as a percentage of set pressure. 'Blow down' is also a term used in steam plant and process plant to describe rapid depressurization by venting (usually gas or vapour).
Blow down valve	A valve, often specially designed, for rapid depressurizing of a high pressure system. Also called a blow-off valve.
Body (valve)	The main pressure-containing component of a valve housing the working components. The body may be of a single piece construction or may be comprised of several segments.
Bolted bonnet	A term specifying that the bonnet is bolted to the body (not screwed or welded).
Bolted gland	A term specifying that the gland is bolted to the bonnet (not screwed or a union type).
Bonnet	The pressure-containing top cover containing the gland. Applies to block, diverter, and control valves. The term 'cover' is used for check valve.
Bonnet assembly	The bonnet and valve operator.
Bonnetless valve	A term for valves having all internal parts inserted into a one-piece body (could be used for some bellows sealed valves).
Bonnet packing	See Gland packing.
Bottom flange	A term used when a pressure-containing plate covers an opening on the bottom of a valve.
Brass to iron	Specifies a valve with a brass closure and iron seat or *vice versa*.
Breakout torque or force	The torque or force required at the valve stem to initially move the closure from the fully closed position.

Bronze trim or bronze mounted	Specifies that trim (internal components), e.g., closure, stem, seat ring are of brass or bronze.
Bubble tight	A typical requirement for manufacturer's production test meaning no visible seat leakage when tested on gas (bubbles of air).
Built-up back pressure	A safety valve outlet pressure due to flow into discharge line, normally expressed as a percentage of the set pressure.
Butterfly valve	A valve having a circular disk closure normally mounted on a shaft. Rotation gives a wing-like movement to close at right angles to the flow.
By-pass valve	A valve to divert flow around or past part of a system through which it normally passes.
Cage trim	Cylinder with openings in the wall which surrounds the valve plug. Used to improve regulation of flow and minimize wear in control and choke valves.
Cavitation	A phenomenon which can occur in partially-closed valves due to a reduction in static pressure at the restriction followed by pressure recovery downstream. If the liquid vapour pressure is reached, vapour-filled cavities form and grow around gas bubbles and impurities. Increases in static pressure downstream cause vapour bubbles to suddenly collapse or implode and can damage the valve trim, body, and so on.
Cavity relief	Relief to prevent pressure build up in valve body cavities due to temperature changes of trapped fluids. Internal relief may be provided by the seat design. External relief may be necessary in some designs; this requires a body mounted relief valve with inlet to the cavity.

Chatter	A safety valve term for rapid reciprocating motion of the closure (disk) contacting the seat.
Check valve	A self-acting valve type allowing forward flow. Prevents reverse flow. Other terms used include back pressure valve, non-return valve, reflux, and retention valve.
Chlorine service	Valves designed for this service.
Choke valve	A control valve for regulating flow and pressure in oil field service. The design is similar to a globe valve, the closure being conical or plug shape seating in a circular orifice. Generally used for drilling and production of oil and gas; also used for water injection.
Clamp gate valve	A gate valve with body and bonnet held together by a 'U' bolt clamp which facilitates stripdown for overhaul.
Class rating	See Rating.
Clean service	A classification used to denote that the process fluid and piping system are essentially clean and free of abrasive, dirt, and so on.
Closing torque or force	Torque or force required at the valve stem for closure onto the final seat position.
Closure member	A term for the operating components regulating the flow or pressure e.g., ball, disk, gate, piston, or plug. Synonymous with obturator.
Cock	A small quarter turn, rotary action valve with a taper plug closure, generally used for low pressure instrumentation, drains, and vents.
Combined stop check valve	See Globe stop and Check valve.
Compact gate valve	A gate valve having compact dimensions, normally complying with API 602 may be supplied with one extended end for threaded or welded attachment (API 606).

Conduit gate valve	A parallel gate valve having a continuous uninterrupted port through the valve when fully open. Also known as a through conduit gate valve. See also Slab gate and Split gate valves.
Control valve	A term for a valve that regulates flow, pressure and temperature. See also Automatic control valve and Manual control valve. Other terms include 'regulating valve' and 'throttling valve'.
Conventional	A term used to describe the basic generic valve type, distinguishing it from specific variations e.g., butterfly valves are conventional or high performance, parallel gate valves are conventional or through conduit, safety valves are conventional or balanced.
Conventional butterfly valve	See Rubber-lined butterfly valve.
Conventional parallel gate valve	A parallel gate valve with a gate or disk closure that does not seal the bottom of the body cavity when open. Also called the regular parallel gate valve.
Conventional safety valve	A safety valve having a bonnet vented to discharge (outlet) from the valve.
Corrosion allowance	An addition to the thickness of pressure-containing components subject to corrosive fluids.
Crawl	A term for the graduated adjustment of set pressure of spring loaded safety valves from below normal to normal after the spring temperature has been raised by discharged fluid.
Cryogenic valve	A valve designed to operate within the temperature range $-50°C$ to $-196°C$. Typical valve types used include ball, butterfly, and globe.

Dashpot

A device for damping the movement of the closure member. Dashpots are used for swing check valves to reduce or eliminate surge in piping systems.

Diaphragm

A flexible disk of metal (convoluted) or a membrane of resilient material (elastomers, polymers – often fabric reinforced) that provides limited movement for valve operation and seating/separation of fluids. Applications include diaphragm check valves, diaphragm control valve actuators, and diaphragm valves.

Diaphragm actuator

A control valve actuator consisting of a diaphragm in a pressurized housing. Usually pneumatic operation with spring return.

Diaphragm check valve

A valve in which the closure is a specially shaped diaphragm allowing forward flow.

Diaphragm valve

A block/control valve utilizing a resilient diaphragm as the closure member. Two types available: weir and straight through type. The valve is also known as a glandless valve (gland can be provided). Used for low pressure applications; limited temperature operating range.

Differential pressure

The difference in pressure between any two points in a piping system. Usually the difference in pressure between the upstream (inlet) and downstream (outlet) of a fully closed valve. Also known as pressure differential.

Dirty service

A classification used to denote that the process fluid and piping system contain particulate which may damage valves unless specially selected for the condition (e.g., conduit gate and metal seated ball valves for oil and gas production).

Discharge coefficient	A safety valve term for the ratio of the measured relieving capacity to the theoretical relieving capacity related to the nominal flow area.
Diverter service	A process requirement to direct flow from one stream to one or more other streams. Several block valves or a single diverter valve, may be used.
Diverter valve	A valve with multiple ports to divert flow in several directions and prevent intermixing. Also called multi-port valves, switching valves and, change-over valves. The term may also be used to describe several valves used in combination for directing flow streams.
Double block	The provision of double isolation either by means of two separate valves installed in series or by means of a single valve having two seats, each of which provides a seal.
Double block and bleed	The provision of double isolation with the additional capability of being able to vent the space between the two isolation points.
Double block valve	A single valve having two seats, each of which provides a seal. The seats may isolate with the pressure differential acting in the same or in opposite directions, depending on the design.
Double block and bleed valve	A single valve incorporating two seats, each of which provides a seal, and a means of venting the space between the seats. The seats may isolate with the pressure differential acting in the same or in opposite directions, depending on the design.
Double disk butterfly valve	An uncommon design having two disks to isolate flow and pressure. Usually provided with cavity bleed for double block and bleed.

Double piston effect	A term used for trunnion mounted ball valves in which the pressure energized seats are designed to seal against cavity pressure as well as line pressure. The floating seat is effectively a piston seating by pressure from either side.
Downstream seated valve	A valve where the upstream line pressure loads the closure member against the downstream seat with differential pressure across the valve. Typical valves include, floating ball valves, and high performance butterfly valves.
Drag valve	A control valve with a perforated cage trim for high pressure drop, low noise application.
Drop tight	A term specifying that droplets of liquid shall not be passed when a valve is closed.
Dual plate check valve	See Split disk check valve.
Eccentric ball or plug valve	A quarter turn, rotary action valve. The closure member is pivoted off centre to provide cam action which urges it against the seat.
End entry ball valve	A ball valve with a single piece body in which the ball is assembled from one end of the bore and held in position by a seat/retainer ring fastened to the body.
Erosion	Damage to the valve/trim caused by high velocity flow of fluids.
Excess flow valve	A valve designed to close automatically when flow exceeds a specified rate.
Expanding wedge gate valve	A gate valve with two parallel faced gates separated by an expanding wedge which loads each gate against its fixed seat.
Face to face dimensions	The dimension from the end face of the inlet port to the end face of the outlet port of a valve or pipe fitting.

Facing	The finish of the gasket contact surface of flanged end valves.
Fire fighting valve	Generally a globe valve specifically for fire hydrants having an outlet for a hose coupling.
Fire safe valve	A misleading and incorrect term used to describe a valve suitable for fire hazard areas – see 'Fire test certified valve'.
Fire test certified valve	A valve certified to have been type tested in accordance with accepted fire test standards.
Fixed seats	Seats fixed to valve body. A term differentiating from valves with floating seat.
Flap valve	A low pressure swing check valve with hinged disk or flap, sometimes leather or rubber faced.
Flash point	The temperature at which a fluid first releases sufficient flammable vapour to ignite in the presence of a small flame or spark.
Flat faced bonnet joint	The jointing surface between a valve body and bonnet when the bonnet is not recessed into the body but seated on a flat surface.
Flexible wedge gate valve	A valve having a slotted gate permitting flexing. Improves seat alignment, reduces break out and closing force (torque) with high pressures and temperatures – see Gate valve.
Floating ball valve	A ball valve having a ball supported by seat rings, and free to move relative to the stem. Also called a seat supported ball valve.
Floating seats	Valve seats having limited axial movement, energized by line pressure to seal against the closure member. A term used to differentiate from valves with fixed seats.
Float valve	An automatic flow control valve of globe or piston type, usually operated by a float mechanism for regulating or maintaining liquid level in a tank.

Flow coefficient C_v	The flow, in US gallons per minute of water at a temperature of 60°F, that will pass through the valve with a pressure loss of one pound per square inch at a specific opening position. Typically used in sizing control valves.
Flutter	Rapid reciprocating or oscillating motion of a disk during which the disk does not contact the seat (applicable to safety and check valves).
Foot valve	A lift or swing check valve with an open inlet for total immersion on a pump suction line. Always fitted with a filter or strainer.
Full bore/port	A valve bore approximately equivalent to the pipe bore. Minimizes pressure drop and facilitates pigging.
Full bore diaphragm valve	Diaphragm valve with straight through full bore. See Diaphragm valve.
Full bore plug valve	Plug valve with a full cylindrical bore through valve as opposed to standard patterns with trapezoidal ports.
Full face gasket	A flat gasket covering the entire surface of parts to be joined.
Full lift safety valve	A valve which opens rapidly to the fully open position when the set point is reached.
Galling	The tendency to seizure of two metallic components in sliding contact, usually where there is insufficient difference in relative hardness.
Gate	The closure member of a gate valve.
Gate valve	A linear action, multi-turn (when hand operated) valve. The closure member is a gate or disk closing against flat seats located transverse to the pipe axis.
Gear operator	A gearbox fitted to a valve for manual or actuated operation to provide a mechanical advantage.

Gland	A flanged or screwed component fastened to the bonnet to compress and retain the gland packing.
Gland follower	An integral or separate gland component in direct contact with the packing.
Glandless valve	A valve not requiring a gland, e.g., a diaphragm valve or pinch valve.
Gland packing	See Packing.
Globe stop and check valve	A valve combining the self-acting operation of a check valve with the manual action of a globe valve. Also called a combination stop check valve.
Globe valve	A linear action, multi-turn (when hand operated) valve with a disk or piston seating on a single flat or shaped seat.
Hammer blow handwheel	A handwheel designed to provide a sudden impact load to initiate opening of a valve where a plain handwheel is inadequate but a geared operator is not justified. Some electric actuators incorporate this feature.
Handwheel	The manually-operated component used to open and close a valve.
Hand valve	A term sometimes used to describe small bore valves of the instrument type.
Hard faced seats	Seats or seating having a hard facing or coating to provide good sealing surfaces resistant to wear, wire drawing, galling, and abrasion. Materials generally used: plating (electroless nickel, chromium); cobalt bearing alloys (stellite, colmonoy); carbides (tungsten, chromium).
Hard seated valve	A valve with primary seating of metallic, ceramic, composition (carbon/metal), or other non-resilient materials to provide hard wearing faces.

(Valve) height	Usually the distance from bore centre line to top of the valve handwheel or the top of a geared operation handwheel or powered actuator. NB the overall height includes the distance below the valve bore centreline to the base of the valve.
High performance butterfly valve	A butterfly valve with the closure disk offset or double offset from the stem polar axis, permitting use at higher pressure than conventional butterfly valves and offering tight sealing. Also called eccentric butterfly valves.
Hose end valve	A valve for utility services which has fittings for connection to a hose.
Inlay	A corrosion-resistant coating of body internal wearing surfaces.
Inlet port	The port connected directly to the upstream pipework.
Inner valve seat	See Seat ring.
Inside screw non-rising stem	A stem design in which the gate rises on a threaded portion of the stem within the valve body and below the gland packing.
Iris valve	A valve design in which the multiple leafed closure member moves towards the centre of the valve bore to seat.
Isolation valve	See Block valve.
Kicker valve	A specific term for the valve used on pig traps to start flow and initiate pig launching and to stop flow on pig reception. The valve is, in effect, a bypass valve.
Knife gate valve	A valve having a narrow parallel gate with a shaped edge for cutting through glutenous media or slurries. Normally limited to low pressure, the gland usually seals directly onto the gate.

Lantern ring	A metallic ring which forms a chamber between upper and lower sets of compression packings in a stuffing box. A port from the bonnet connects with the lantern ring which may have several functions: (i) an injection chamber for gland sealants; (ii) a purging chamber for preventing external loss of process media; (iii) a leakage collection chamber.
Lapping	The process of rubbing and polishing surfaces (e.g., closure and seats) to obtain a smooth seating surface to minimize leakage.
Leak tight	A term confirming there is no leakage through a valve.
Lever operated valve	A quarter turn, rotary action valve, e.g., ball, butterfly, and plug operated by a lever or 'T'-bar. Usually small, low rated valves.
Lift check valve	A check valve in which the closure member is lifted from the seat during forward flow, e.g., ball, piston, nozzle check, and foot valves.
Lifting plug valve	A valve in which the plug is lifted clear of the seat during rotation from open to close to reduce operating torque. Purging of the exposed cavity by steam or other fluid may be needed for some severe service.
Line blind valve	A block using a plate arranged to rotate between flanges that are clamped against the plate to make a seal. Also called a goggle and spectacle valve.
Lined valve	A valve (usually plug, ball, diaphragm) with the body internally lined, (typically with PTFE).
Line pressure	The pressure in the piping system.
Lubricated plug valve	A plug valve having grooved seating areas and passageways for lubricant injection

which reduces operating torque and assists sealing. See Pressure balanced plug valve.

Lugged pattern
Term for butterfly valve of wafer pattern provided with lugs drilled to accept flange bolting. Other terms are lugged wafer pattern and single flanged valve.

Manual control valve
A valve for manually regulating flow or pressure either directly or indirectly e.g., powered actuator. Globe, needle, ball, butterfly, and diaphragm valves are extensively used.

Manual operation
A valve supplied for manual operation by lever, handwheel, or endless chain.

Manual override
Operation by handwheel of a powered actuator in the event of power failure and for setting of the actuator travel in relation to the valve.

Mechanically seated valve
A valve designed to achieve primary seating by mechanical means e.g., expanding wedge gate valve, with a (wedge gate) valve and the expanding plug valve.

Mixing valve
A control valve, usually self-operating, which uses a control element to regulate flow input(s). Multi-port designs e.g., globe type may be used to mix different fluids and are sometimes called blending valves. Mixing valves are also used for temperature control in conjunction with a thermostat.

Modulating safety valve
A valve designed to modulate between open and closed over the entire or a substantial portion of the valve lift.

Needle valve
A globe valve with a conical plug (needle) closing into a small seat. Used for flow metering and damping pressure fluctuations on instruments e.g., pressure gauges. Instrument isolating valves are often called needle valves.

Nominal bore/nominal size	A number denoting the basic size of a pipe, fitting, or valve. The actual bore may vary considerably from the nominal size.
Non-lubricated plug valve	A plug valve not depending on the injection of lubricant or sealant for operation and seating e.g., lined/sleeved plug valves and lifting plug valves.
Non-rising stem	See Inside screw, non-rising stem.
Nozzle check valve	See Axial piston valve.
Oblique pattern	A valve body with the closure member at an angle to the port reducing pressure drop. Used for linear action globe and lift check valves. Also called a 'Y' pattern valve.
Obturator	The closure member of a valve; the moving component which seals against the seat.
Outlet port	The port connected directly to the down-stream pipework.
Outside screw and yoke (OS&Y)	A rising stem design for linear action valves (gate and globe). The threaded portion of the stem is isolated from the process media by the gland packing and the stem rises via a threaded nut in the yoke.
Outside screw, rising stem	See Outside screw and yoke.
Overlay	A hard facing of trim components (welded or sprayed overlays as opposed to coatings and platings).
Overpressure	A safety valve term for pressure increase above the set pressure (expressed as percentage of set pressure) permitted during discharge.
Packing	The material used to seal the valve stem within the valve body bonnet.
Packing assembly	The gland, gland follower, bolting, and packing components of a valve.
Packing box assembly	The bonnet gland stuffing box and packing assembly used to seal the stem.

Packing nut	A nut that is used to directly compress the packing in the valve stuffing box.
Parallel gate valve	A gate valve with a parallel slide gate or disk closure. See Conventional parallel gate, Conduit parallel gate, Slab gate and Split gate valves.
Parallel plug valve	A valve with a cylindrical plug closure, various methods of sealing are used i.e., lubricated, expanding plug, eccentric, and 'O' ring sealed designs.
Parallel slide valve	A parallel gate valve with a spring-loaded double disk closure (see BS 5157). Generally used for power generation and steam service applications.
Particulate	Small solid contaminants in the process media which may be abrasive and damaging to valve performance.
Penstock valve	A gate valve with a rectangular gate mounted in a frame fixed to a wall or bulkhead. Used for handling large volumes of water. Also called a sluice valve.
Pig-ball valve	A ball valve designed to hold and launch or receive pigs or spheres.
Piggable valve	A full bore block valve suitable for the passage of pigs and spheres. Valves used are trunnion mounted ball valves, conduit gate valves of slab, and split wedge design.
Pilot operated safety valve	A valve comprising of a main valve and a pilot valve where the load on the main valve disk is applied by fluid pressure. The pilot valve senses the pressure of the system and causes operation of the main valve through its full travel.
Pilot operated safety valve with restricted loading	A pilot-operated safety valve in which the closure seat loading is restricted to permit the valve to fully open within the permissible

	overpressure should the pilot fail to remove the seat loading.
Pinch valve	A valve comprising a flexible tube, either exposed or enclosed in a body. The tube is pinched to close mechanically or by application of fluid pressure in the body.
Piston check valve	A check valve with a free or spring loaded piston type closure member.
Piston valve	A globe valve in which a piston closure enters or withdraws from a seat bore to start, stop, or regulate flow. The seat bore or piston contains packing to effect a seal. In addition to the standard straight pattern design other variants are available, e.g., tank bottom outlet valves and sampling valves.
Plate check valve	A valve with several metallic membranes which allow forward flow, but close to prevent reverse flow. Used for air and gas compressors.
Plug	A valve closure member which may be tapered or cylindrical in shape.
Plug valve	A quarter turn, rotating action valve where the closure member is a plug sealing against a downstream seat.
Poppet valve	A linear action valve in which the closure is a stem mounted disk often with an angled edge closing on an angled seat.
Port	A passage through a component, e.g., the inlet and outlet through an open valve. Also used to denote the valve seat opening.
Preferred flow direction	A term for a design of valve in which line pressure aids closure in one direction and tends to unseat from the other direction. Valves with preferred flow directions include high performance butterfly, eccentric ball, plug and split (parallel) gate valves.

Pressure balanced plug valve	A lubricated valve with internal passageways designed to reduce the unbalance pressure load on the plug.
Pressure control valve	An automatic valve with a pressure sensing element to control pressure within set limits (see Pressure maintaining valve).
Pressure differential	See Differential pressure.
Pressure energised seat	A floating seat with limited movement that is energized by line pressure which provides the seating load against the closure member. Used for trunnion mounted ball valves and slab (parallel) gate valves.
Pressure maintaining valve	A control valve that maintains the upstream line pressure within specified limits. Also called a pressure sustaining or pressure control valve.
Pressure reducing valve or pressure regulator	An automatic control valve with an internal pressure sensing element that reduces the upstream line pressure to a set level downstream and maintains the reduced pressure, irrespective of changes in upstream pressure (see Back pressure controller).
Pressure relief valve	Traditionally a safety valve in which the degree of opening is proportional to pressure. Used on liquid service only.
Pressure sealed valve	A valve in which the bonnet is in the form of a cover plate held in place by segmented ring sections and sealed by a taper ring gasket that increases the sealing effect as internal pressure is increased. This design replaces the bonnet and body flanges with the benefit of saving weight and cost.
Pressure/vacuum safety valve	A valve that automatically releases excess pressure or admits pressure to prevent a vacuum forming when filling or emptying a pipeline or tank.

Quick operation gate valve See Lever gate valve.

Raised face flange (RF flange)	A pipe flange or valve body flange in which the gasket contact face is raised proud of the flange face.
Raised face gasket (RF gasket)	A gasket with dimensions matching a RF flange gasket face.
Rating	A designation (e.g., class 300, PN40) which defines the combinations of pressure and temperature at which a valve made from a particular material may be used.
Reduced bore	A valve in which the diameter of the valve bore is reduced in the central portion of the body. Generally the reduction is to the next pipe size down but may be less. Reduced bore valves are used to save weight and cost where the additional pressure drop can be tolerated (see Venturi valve).
Regular pattern	An American term for a valve design in common usage as opposed to less common designs. Also specifically used for plug valves in which the port commonly used is trapezoidal and equivalent in area to a circular bore. Also known as a standard pattern (see Conventional design).
Relief valve	See Pressure relief valve. In the USA a direct acting valve where the lift is proportional to the amount by which pressure exceeds the set point. Liquid service only.
Reseating pressure	The pressure at which a safety valve reseats after discharge.
Resistance coefficient	A coefficient defining the friction loss through a valve in terms of velocity head or velocity pressure.
Reverse acting gate	A term used for conduit gate valves in which the gate(s) is ported such that the gate is

raised to close. This design is used to minimize ingress of particulate into the body cavity where the process fluid cavity is dirty. It also utilizes the unbalanced pressure force on the stem to assist closure.

Ring type joint flange (RTJ flange)
A pipe or valve body flange with a machined groove into which a metal ring type joint is fitted.

Ring-type joint gasket
A metal ring of oval or octagonal section.

Rising stem
See Outside screw, rising stem, Outside screw and yoke and Inside screw, rising stem.

Rotary disk valve
A quarter turn development of the parallel slide valve design in which an offset shaft slides a disk across the valve seat to open or close. Designs include a single disk closure and a double disk closure in which the disks are separated by a spring. Upstream pressure loads the downstream disk against its seat. Also called a rotary gate valve and a lever gate valve when manually operated by a lever.

Rotary valve
A general term sometimes used to describe valves with a rotary action, e.g., ball, butterfly, plug valves.

The term is also used specifically for a design of valve used to dose precise quantities of powders or liquids into a process stream. The closure is machined with a series of cups to hold the dose and may be continuously rotated through 360 degrees or partially rotated to provide dosing. A particular design is the cup ball valve typically used in chemical plants.

Rubber lined butterfly valve
A quarter turn rotary action valve of conventional design in which the rotational axis of the disk is concentric with the shaft axis. The rubber lining of the body forms the valve

	seat. The lining may be vulcanized to the body or be replaceable. Valves of this type are limited to low pressure and temperature applications.
Rubber seated wedge gate valve	A valve with a rubber coated wedge closure that seals against the valve port. This design eliminates the cavity in the bottom of the valve but application is limited by pressure and temperature.
Running torque or force	The torque or force required to stroke a valve over the majority of its travel. This is always less than the breakout and closing torque/force.
Safety valve	Nowadays a term for any automatic valve that relieves a pressure system where abnormal operating conditions cause the pressure to exceed a set limit and which closes when pressure falls below the set limit. In the USA it is specifically a full lift pressure-relief valve intended for gas or steam service.
Safety relief valve	A term now obsolescent in the UK. See Safety valve. In the USA it is specifically a direct acting pressure-relief valve intended for gas, vapour, and liquid service.
Sampling valve	A globe valve designed for taking samples from a vessel or piping system. Usually mounted on the bottom of a vessel or pipe (see Piston valve).
Screw bonnet	A bonnet screwed into the valve body rather than flanged or bolted. Generally used on small bore low pressure valves for non-hazardous industrial applications and for instrument isolation. The bonnet is sometimes seal welded to prevent external leakage from the threaded portion.

Screwed end	A valve (and pipe) with ends threaded, suitable for a screwed connection.
Screwed gland	A gland usually threaded internally and screwed into the bonnet stuffing box, rather than flanged and bolted. Generally used on small bore valve, e.g., instrument valves. A locking device is sometimes fitted to prevent rotation in service due to, for example, vibration.
Sealant injection	The injection through a non-return valve of a sealing compound to seal the gland and/or seats of a valve either by design (lubricated plug valve) or for use in an emergency (leaking valve).
Seat	The part of a valve against which the closure member bears to provide shut-off and through which process fluid flows. May comprise one or more compounds of metallic or non-metallic materials.
Seat bush	See Seat ring.
Seat holder	A metallic component, usually corrosion-resistant, with an insert of another material, e.g., a soft seat of elastomer or polymer. Specifically the term is used for valves with floating seat designs. The seat holder is tubular in shape permitting axial movement in a housing and line pressure against the closure member. Typically used in trunnion mounted ball valves and slab (parallel) gate valves.
Seat housing	The counter bore in the body of floating seat valves in which the seat holder is located.
Seat insert	A ring shaped element generally of soft material such as elastomer or polymer that is fitted in a seat holder, seat ring or closure. Also called an insert seat and soft seat insert.

Seat pocket	A general term for the counterbore within a valve body that houses a seat holder or seat ring. Seat pockets may be left plain or overlaid in corrosion-resistant materials.
Seat ring	A ring-shaped metallic component, usually corrosion-resistant. One face forms the valve seat which may contain a seat insert. The reverse end is fastened to the body of the valve by a threaded portion or by swaging in small bore valves. Seat rings are typically used in valves of fixed seat design where seat replacement is required, although seal welding is sometimes employed.
Seat sealing pressure	The pressure required to prevent leakage across the seating contact surface of a valve seat and closure member. Sealing pressure may result from line pressure and/or mechanical force and is dependent on the seat material, surface finish and geometry of the valve trim. Also called seat contact pressure.
Seat supported ball valve	See Floating ball valve.
Seat to body seal	The seal preventing leakage from a valve body cavity to the line via the seat-to-body connecting interface. In fixed seat valves the seal usually depends on a screwed or swaged joint. In floating seat valves a dynamic seal is required to allow float of the seat assembly. This seal may be in the form of an 'O' ring(s), lipseal, metal spring seal, or bellows.
Set pressure	The pressure at which a safety valve is designed to open.
Shaft	See Stem. Usually associated with rotary valves.
Short pattern valve	Valve designed with shorter face-to-face dimensions than normal. Several standards

	include short pattern dimensions for different types of valves. Generally, valves are of reduced bore.
Shut-off valve	See Block valves.
Side entry ball valve	A two or three piece body ball valve in which the ball is assembled via the bore of one of the body pieces.
Single flanged valve	Normally a wafer pattern valve provided with a single flange or lugs for bolting to pipe flanges.
Single piece ball valve	A ball valve with the body in one piece, the ball being assembled from one end and held by a retainer ring which may also include a valve seat. Generally used for small bore valves.
Size	A number defining the dimensions of a valve e.g., 6″ NB DN 150. Sometimes the actual internal diameter is given, typically for well-head gate valves. Reduced bore/venturi valves normally reference the size of the reduced nominal internal diameter e.g., 6 in × 4 in. NB (see also Full bore/port and Reduced bore).
Slab gate valve	A conduit gate valve with a single parallel sided gate normally sealed by floating seats. Line pressure energized with supplementary mechanical loading (springs) for low pressure. Fixed seats are sometimes used. Generally for small bore valves.
Slam shut valve	A valve having a disk which is held open against the flow and closes automatically when released. Fulfills similar safety function to excess flow valve but requires external signal to initiate operation.
Sleeved-plug valve	As plug valve having a sleeve of PTFE or similar polymer material machined and press fitted (or keyed) into the valve body.

Slide valve	A general term for the knife gate valve and the parallel slide valve. Specifically, the term is used for parallel gate valves of the slab design working at high temperature, low pressure, and equipped with purging facilities. Used for gas/catalyst service on cat cracking units in refineries.
Slurry	A product consisting of solids mixed with liquid to assist transportation through piping systems. The solids vary widely and may be abrasive, non-abrasive, hard or fibrous material.
Small bore valve	A term generally used for valves of DN 50 and less.
Socket weld end	A valve with counterbored end(s) into which piping is fitted and fillet welded to make a joint.
Soft seat insert	See Seat insert.
Soft seated valve	A valve with primary seats manufactured from elastomers, polymers, and similar re-silient or semi-resilient materials.
Soft seated wedge gate valve	A wedge gate valve in which primary soft seating is used. Seating may be a solid PTFE seat or a soft seat insert, either in the valve body or in the wedge gate closure.
Solenoid valve	A linear action block valve, generally of globe type, fitted with a solenoid for quick operation. Hermetically-sealed designs available. Usually DN50 or smaller.
Solid wedge disk	A disk or wedge of one piece.
SP	An abbreviation for steam pressure. A number following the abbreviation is the maximum non-shock operating pressure in psi at a given temperature.
Split wedge gate	Gate valve having a wedge, comprising two parts in order to provide flexibility and resist-ance to thermal wedging.

Stem	The component of a valve which transmits force or torque from the operator to the closure member.
Stop check or screw down non-return valve	A valve which automatically closes when flow reverses and which can be screwed down into a stop or closed position.
Stuffing box	The gland packing chamber within a valve bonnet.
Swing check valve	A valve which has a swinging disk that opens automatically when flow is established and closes automatically when flow ceases or is reversed.
'T' bar	A lever handle extending on both sides of the stem/shaft centre line.
Throttling	Regulation of flow through a valve.
Trim	Internal components associated with isolating or regulating the flow. Includes seating surfaces, closure members (gate, disk, ball, plug, etc.) stem, bearings, guides, and associated parts.
Uni-direction valve	A valve designed to seal in one flow direction.
Upstream seated valve	A valve which, when closed, seals by line pressure on the upstream seat.
Valve closure member	That part of a valve which is positioned to close, open, or to control the amount of flow. Synonymous with obturator.
Valve operator	The valve part or parts through which a force is applied to move or position the valve closure member onto the seat.
Valve operator, manual	A valve operator consisting of a hand lever, wheel, or other manual device.
Valve operator, mechanical	A valve operator consisting of a cam, lever, roller, screw, spring, stem gearbox, or other mechanical device.

Venturi throat valve	A valve having a reduced port opening at the seat with the body tapered to the ends to produce a venturi effect. Minimizes the velocity head losses associated with the reduced port.
'V' port plug or ball	A type of valve closure member (plug or ball) having a 'V' sheared orifice; it has good throttling characteristics.
Wiper ring	A ring which removes debris etc. by a wiping action.
Yoke	That part of a valve which connects the valve actuator or operating mechanism to the valve body in rising stem valves. Sometimes an integral part of the valve bonnet.
Yoke bush	A screwed bush, retained in the yoke (often in bearings) to which the handwheel or actuator is connected. Rotation of the bush causes valve stem to move up or down.
'Y' type globe valve	See Oblique pattern.

BIBLIOGRAPHY

These references may be useful in providing further information on valve application, sizing, and design. It is possible that some may be out of print.

J. W. Hutchinson (editor) *ISA Handbook of Control Valves*, Second edition, 1976 (Instrument Society of America).

Good information on valve sizing for various flow conditions as well as many other aspects of control valves.

D. S. Miller *Internal Flow Systems*, 1990 (Gulf Publishing Company).

Information on flow resistance coefficients for isolating valves in various flow conditions.

J. L. Lyons (editor) *Lyons' Valve Designers' Handbook*, 1982 (Van Nostrand Reinhold, New York).

Contains wide-ranging articles on many aspects of valve design.

R. W. Zappe *Valve Selection Handbook*, 1991 (Gulf Publishing Corporation, Houston, Texas).

Covers selection of isolating valves, relief valves, and rupture disks.

G. H. Pearson *Valve Design*, 1979 (Mechanical Engineering Publications Limited, London).

Now somewhat out-dated, but gives useful insight into design requirements for common valve types.

D. R. Airey (editor) *Valve and Actuators Applications and Developments*, 1990 (Independent Technical Conferences, Kempston).

Issued Bi-annually. Contains papers on current developments in valve and actuator technology.

R. C. Whitehouse *The Valve and Actuator User's Manual*, 1993 (Mechanical Engineering Publications, London).

Manufacturers' publications give useful information about valves, including sizing and materials. Crane have, for many years, published an excellent book called *Flow of fluids*, and the publications of control valve manufacturers are usually worth seeking out.

INDEX

193

Printed and bound by CPI Group (UK) Ltd, Croydon, CR0 4YY